Single women, land and livelihood vulnerability in a communal area in Zimbabwe

Single women, land and livelihood vulnerability in a communal area in Zimbabwe

AWLAE SERIES No. 9

Gaynor Gamuchirai Makura-Paradza

Wageningen Academic Publishers

ISBN: 978-90-8686-146-0
ISBN: 978-90-8686-700-4
DOI: 10.3921/978-90-8686-700-4

ISSN: 1574-2474

Cover drawings: Veneka Paradza

First published, 2010

This thesis is dedicated to the memories of
my mum
Eunice Makura
&
my sister
Gennis Varaidzo
You would have been so proud

Acknowledgements

If I mentioned all the people who assisted me with this thesis, there would not be enough space to present my findings. To all those who supported and encouraged and assisted me along the way, I am grateful. The people who facilitated this work by agreeing to become the subjects of my study in Goromonzi District and at the Widows and Orphans' Trust in Harare Zimbabwe *Ndinotenda.* I am deeply indebted to the 22 women who generously gave their time and extended their hospitality to allow me to examine their difficult lives while at the same time expecting nothing tangible in return. The supervising Professors Han van Dijk, Bernd van der Meulen and Ab van Eldjik who gave me invaluable support. Han, somewhere in 2006, I stopped counting the number of drafts you had read. I admire you for your patience, guidance and support which extended beyond Wageningen and academia. I am indebted to Bridget O'laughlin for her critical comments which continued into your sick leave well into your retirement. I thank the editor, Catherine O'Dea for the professional editing of my work. I also thank Bridget and Han for being patient and encouraging me to take time out of the PhD when 'life' got in the way. Julie Stewart, I appreciate the opportunities I had to walk into your office and have discussions during my visits to Zimbabwe.

To all the staff in Law and Governance Chair group, I thank you for all your support and cards and emails that you send to me individually and collectively while I was in Zimbabwe. I would like to extend a big thank you to the AWLAE team in the Netherlands and Nairobi especially Margaret, Norah and Carolyne and the late Julia Gitobu. I owe a debt of gratitude to Hedy Munro, Riet Westering, Ellen Wegkamp and Maarit Junnikkala at Wageningen University for supporting me throughout my research. Lisa Price and Anke Niehof, I appreciate your concern for my welfare and the interventions you made on my behalf throughout this process.

To the African Women Leaders in Agriculture and Environment members, it was a privilege to meet and work with you. AWLAE touched my lives in many different ways that I will treasure for the rest of my life. To my housemates at Van Uvenweg, Doris Kakuru, Kidist, Mariame Maiga, Corrie Du-Preez, Nakar and Faith Nguthi I will always cherish the memories we shared. Anna Szajkowska, I thank you especially for keeping my spirit up by continuously looking up to me (*even when I seemed not know where I was going with the PhD*). If I did this, so can you. Thank you for being such a wonderful friend and office mate. Doris. I thank you for constantly assuring me that this day will come when I had no reason to share that vision

The Wageningen Community, the Student Chaplaincy, Amazing Grace, Zimbabwe Community, Carijn Vijfhuizen, Anesu Zvinavashe and Anotida Maphosa, Han and his wife Mirjam and their children, I thank you for the invites to your house and for the privilege of sharing in your growth which constantly reminded me of my children who I missed terribly. I thank my friends who provided support for me in ways too numerous to mention on this page. Bertha Chiroro, Dr Doris Kakuru, Beatrice Gwata, Ali Stangroom, Anna, Stella Muchemenyi Naomi Kembo, Cordelia Muchatuta, Mrs Chirumhanzu, Memory Mawema, Dorcas Zireva, Tarisai Zinyuke, Beatrice and Gwata, Meagan George and Stewart Gordon, Lindiwe Sibanda and Jeanette Mazonde, Petronella Chaminuka, Lebogang Mokwena, Monica Bingudza, Loveness Zimuto and the 'Congress and Associates' Lindiwe Sibanda and Jeanette I am eternally indebted.

Janice and Lenny Gordon, Nyasha Mazonde, Katrina Stangroom, Zuva and Tafara Gwata, Kudzai and Rutendo Furusa, Angelica and Dylan Sibanda, Kombo Bingudza, Thabo, Thandeka and, Thando Muchemenyi, Nakai and Tawana Makura who shared their homes with my children and whose parents supported my family to ensure that my children continued to have a life while I worked on this thesis, *ndinotenda*.

I would like to thank Nelia Hora for more than a decade of faithful service. God bless you always. I would like to extend my gratitude to the Paradza family. I thank Sekuru Paradza for allowing me to draw into his extensive experience in agricultural extension to inform my work. I also extend my thanks to my sister Gracelda for moral support and being my main cheerleader through long calls. Stewart, Moky Nakai and Tawana and family my brother for constant support with your dreadful '*how is the thesis panning out?* You also contributed invaluably through your calls and incentives to Panashe. Stewart, I do hope you read this thesis! Gillian, my sister no words can describe the gratitude I have for your selfless acts towards my children and family. You gave up your own life to be the mother of my children. I will always remember with fondness Veneka's constant references to me as 'auntie' which bear testimony to the close contact you had with my children. The memories of my late sister Gennis Varaidzo continued to be a source of encouragement. Were you here today Gennis, I know you would have been so proud!! Uncle Arthur and his family I thank for the support and mentoring.

I thank you Panashe my son for assuming cooking responsibilities, editing the thesis, excelling in your studies and being a good role model for your sister. Veneka your good start at primary school was a source of encouragement. I will always appreciate the diligence you put into designing the cover for this thesis. I thank my husband Joe for the support you rendered throughout the study period. I thank you for holding the family together, believing in me when I despaired talking me through my infinite resolutions to quit this PhD.

I would like to thank my parents for raising me to be a hardworking person and for giving us the values we have. My father Nyikadzino and my late mum Eunice Makura, I thank God for the years he gave us to be together and for the guidance you provided through the values you instilled in your home. I also thank my late grandfather who encouraged me through his own unique ways. I thank God for making this day possible.

Table of contents

List of abbreviations

AGRITEX	Agricultural and Technical Extension
AIDS	Acquired Immune-Deficiency Syndrome
CSO	Central Statistic Office
ESAP	Economic Structural Adjustment Programme
FAO	Food and Agriculture Organisation
GIS	Geographical Information System
HIV	Human Immuno-Deficiency Virus
IDS	Institute of Development Studies, University of Zimbabwe
IFPRI	International Food Policy Research Institute
GMB	Grain Marketing Board
MDC	Movement for Democratic Change
NAC	National Aids Council
NGO	Non-Governmental Organisation
TB	Tuberculosis
UNDP	United Nations Development Programme
VIDCO	Village Development Committee
WADCO	Ward Development Committee
WLSA	Women and Law in Southern Africa
ZANU-PF	Zimbabwe African National Union (Patriotic Front)
ZHDR	Zimbabwe Human Development Report
ZWOT	Zimbabwe Widows and Orphans Trust

List of tables

List of boxes

List of illustrations

Glossary of Shona terms

Chigara mapfihwa	The practice whereby a deceased woman's sister moves in with the widower as a replacement for the wife
Dongo (singular)/ *Matongo* (plural)	Vacant but previously inhabited homesteads and/or agricultural land
Kugarwa nhaka	The practice whereby a widowed woman is looked after by the brother of the deceased husband
Kutema ugariri	The practice whereby a poor man who is not able to pay bride wealth lives and works at the wife's homestead providing his labour in lieu of bride wealth
Kutizira	Elopement; a girl moves to the homestead of her future husband without the consent of the approved channels
Kuzvarira/Kuroodza/ Kwendisa	Poor families in difficulties betroth a young daughter to a family that will provide bride wealth, which enables the family to survive
Mambo	Chief
Mapoto	A type of a co-habitation; this union is not based on male dominance but on negotiated relations between a man and a woman. Since there is no bride wealth, *mapoto* wives do not have the obligations of women in patriarchal marriages, such as having to move to the man's village
Matorwa	The girl is collected from her parents' homestead by the husband's aunt
Maricho	Piece work performed for other villagers in exchange for food, soap, salt, old clothing or money
Muchinda	Second in command to the chief; also referred to as sub-chief
Musenga bere	The practice whereby a man rapes a girl in order to get her to marry him. This is a forced marriage and is forbidden by the government
Ngozi	*Ngozi* is believed to be the avenging spirit of a person who has been killed. The family suspected of a killing has to give a daughter to the family whose member was killed. That woman will be married to a male relative of the deceased to bear children
Roora/Lobola	The family of the man makes payment to the family of the woman in the process of marriage
Sabhuku (singular)/ *Masabhuku* (plural)	Village headman
Sadza	Zimbabwe's food staple made from ground maize
Zunde raMambo	*Zunde raMambo* (chief's granary) is a pre-colonial traditional social security arrangement designed to address the contingency of drought or famine

Chapter 1
Livelihood vulnerability of single women in rural Zimbabwe

1.1 Introduction: single women in communal farming areas

This research focuses on the ways in which single women access residential and agricultural land and make livelihood decisions in the communal[1] areas of Zimbabwe. Since the establishment of communal areas, there have been significant changes in the socio-economic and political environment of the country. The changes caused by the liberation war, independence, structural adjustment programmes, land reform and redistribution, hyperinflation, the HIV/AIDS epidemic and operation Restore Order (see Section 3.2.4) have induced fundamental changes in the way people make livelihood decisions and access land. The impact on communal areas has been devastating as the productive capacity of these areas has deteriorated to the point where their inhabitants have come to rely on subsidies and transfers from the other sectors of the economy. Single women are probably more adversely affected, as they are not entitled to land under customary tenure rules and form an increasing part of the communal area population. This chapter reviews the theoretical debates underlying the question of women's access to land in Zimbabwe and outlines the concepts needed to understand the choices single women have to make to confront adversity and the decisions they take. This research was initially designed to document the ways in which gender inequality in relation to control of land made single women particularly vulnerable to impoverishment in Zimbabwe. As the research progressed however, it became obvious that single women's land access to land was much greater than the literature suggested, and more complex than the distinction between primary and secondary traditional land rights allows one to see. The question of gendered access to land is embedded in historical changes in land tenure and resource governance, and in the nature of rural livelihoods.

Much of the literature on the impact of AIDS in Africa emphasises the vulnerability of widows arising from women's secondary rights to land (Andersson 1999, Hindin 2002, Walker 2002). The literature, which focuses on land tenure and property relations, emphasises the fact that single women are in a disadvantaged position through patriarchal systems of succession to land. It looks at the end of women's marital relationships, through death of the male spouse or divorce. Single women (widowed, divorced, never married) are seen as a deviation from the norm. This view leads to the misplaced view that the importance of land derives only from its

[1] Communal Lands is the post-independence name for the former Native Reserves or Tribal Trust Lands (Coudere and Marijse, 1991:70).

utility as an agricultural asset and that women access land only as dependents of men (as daughters, sisters and wives). This thesis recognises that singleness is not a transient phase for many women in rural Zimbabwe. We must look not only at dispossession, but also at how single women obtain productive resources and use them to construct rural livelihoods. The dispossession narrative also focuses too narrowly on agriculture as the main livelihood activity and treats rural communities as agrarian.

Such assumptions cannot be maintained in view of all the changes in land access and the vulnerability of the livelihoods of single women in the communal areas in Zimbabwe. Marriage has for some time ceased to be the universal basis of the domestic unit in rural Zimbabwe (Adams, 1991a; Pankhurst, 1991).Women construct and maintain their own livelihoods independent of men and without holding recognised property rights to land. As a result of the collapse of government support for agriculture, and the lack of capital and labour, agriculture is no longer the mainstay of livelihoods in communal farming areas. The rapid changes in communal area livelihoods require that the way in which we assess viability of rural livelihoods be reconsidered as the composition of domestic units and their strategies have to be constantly adapted in view of new and often unpredictable conditions. The next section explains how we do this.

Instead of focusing only on customary land tenure and property relations, this research focuses on a broader concept of land access to analyse the tenure position of women. The research looks at decision making by single women in face of the multiple challenges emanating from unpredictable economic, political and social conditions, and rapid changes in domestic units. The livelihood decision making of single women is analysed through the concepts of access, homestead and hearth-holds, which put women at the centre of the research. The pathways concept (de Bruijn and van Dijk, 2005) is used to analyse livelihood decision making and to illustrate the diversity of livelihood activities and portfolios of single women. This approach was developed to study decision making in risky areas, and focuses on the improvised character of many livelihood decisions and is therefore a more meaningful tool as it enables focus on single women's changing livelihoods and events over their life course.

In the context of the extreme economic crisis, and high and increasing AIDS mortality, rural women heading their own households (particularly widows) are thought to be particularly vulnerable. This vulnerability was thought to be contingent upon their secondary land rights (Bruce, 1988; Cousins, 1990; Maboreke cited in Cousins, 1993; WLSA, 1996). However, the fact that women do not have property rights under existing tenure relations in communal farming areas does not mean that they do not have access to land. The vulnerability of single women and the ways they deal with it through their own capacities must be understood in a much broader context that reflects the various processes of politics, market and histories of communities that affect single women. Single women's agency and livelihood decision making must be understood within the broader framework of the historical evolution of the communal areas and Zimbabwe as a whole.

Communal areas historically functioned as residence, fallback and food production areas for the majority of Zimbabwean Africans. African livelihoods were believed to be based on patrilineal structuring of land, agricultural inputs and natural resource access and governance in communal areas. This dominant view on communal farming areas presents men as having primary rights to land and women secondary rights as daughters, sisters and wives (Cheater, 1986; Berkvens, 1997; Mvududu and McFadden 2001:110). In such systems, adult women can only access communal area resources through marriage or through relations with their agnatic kin group. This made marriage the most important relationship for adult women living in communal areas. Historically within the marriage institution, tasks were divided between the husband and wife to ensure the sustainability of livelihoods of the conjugal unit. The main livelihood activity in communal areas was subsistence agriculture subsidised by wages earned from urban areas. Women (wives) were responsible for the former, and men (husbands) were responsible for bringing in the cash to subsidise these activities (Potts 1995, 2000a). As a result, women were based in communal areas and men were migrant labourers. Land was viewed as the central resource in a largely agro-centric and male-centred livelihood system (Pankhurst, 1991). This dominant view needs to be re-assessed in light of the recent and fundamental changes in the communal farming areas that have led to an increase in single women.

The changes include the migrant labour system, which left many women on their own and vulnerable. The economic crisis and recession undermined men's ability to marry and sustain families. The decline in polygamy due to the influence of the church[2], the increasing economic independence of women following the attainment of independence by the country in 1980 and HIV/AIDS were some of the changes that altered conditions in the communal farming areas (Scoones *et al.*, 1996; Potts, 2000a,b; Francis, 2002; Goebel, 2005). The increased infant mortality, barrenness, infidelity, insubordination, disenchantment with patriarchal marital obligations and moral dislocation, disagreement over migrant remittances between a man's wife and his mother, and changes in matrimonial legislation caused an increase in the number of divorced women (WLSA, 1994). Divorce rates increased in rural areas of Zimbabwe because of spousal abuse and economic hardship following structural adjustment programmes leaving many men unemployed. The desperate need for cash under structural adjustment changed the nature of bride wealth from a socially bonding transaction into an economic one. In the past, one effectively had

[2] 'Church' as used in this thesis refers to the various denominations of Christian churches working in the study area at the time of the research.

a lifetime to pay bride wealth but now, if the son-in-law does not pay up under an accelerated schedule, the father takes his daughter and family back home (Kinsey cited in IFPRI, 1995; Jackson, 2007).

The HIV/AIDS epidemic has also contributed to increasing the number of single women. The epidemic has not just increased mortality, but also articulated on-going processes of social and economic change. Under the threat of AIDS, women leave marriages when their husband's behaviour threatens the lives of the rest of the family (Kinsey cited in IFPRI, 1995). The gender asymmetry in infection rates for HIV/AIDS initially resulted in more adult male deaths than female (Gregson *et al.*, 1997; UNAIDS, 1998). HIV/AIDS has also increased the numbers of single women, as men are reluctant to marry women whose spouses are suspected to have died from the disease (Kinsey, 1995 cited in IFPRI, 1995; Vijfhuizen, 2002; Muzvidziwa, 2002; Jackson, 2007). The number of widows has also increased because of the higher mortality rates among men who are usually older than their wives in Zimbabwe (Jacobs, 2002; Chimhowu and Hulme, 2006:734). As a result of all these tendencies, adult women who do not form an economic unit with a man have become an important category in communal farming areas. These women fall outside the dominant discourse of gendered resource access.

The research context – Zimbabwean communal areas – has been undergoing other rapid changes in the past two decades as a result of the attainment of political independence, economic structural adjustment programmes, land reform, HIV/AIDS, economic meltdown and operation Restore Order (Potts, 1995; Scoones and Wolmer, 2003; Hartnack, 2005). These changes have increased the vulnerability of communal area livelihoods by undermining the communal area economy. The people's reactions to the increasing risks have induced changes in the nature and character of relationships governing resource access in the communal areas. Initially, this study sought to assess the impact of high AIDS-induced mortality on single women's access to land in these areas. AIDS morbidity and mortality put pressure on women's capacity to farm and, it was believed, compromised their land rights that were thought to be derived from men's primary rights (Adams, 1991a, b; Berkvens, 1997; Walker, 2002). However, during the fieldwork it was found that many single women live in communal areas despite the fact that they do not have primary rights to land. In many instances, uncultivated land has been allocated to these women. An investigation to understand the impact of HIV/AIDS mortality on women's access therefore has to take account of other variables that impinge on women's access to residential and agricultural land in communal areas.

1.2 Understanding land rights and gender in Zimbabwe's communal areas

The complexity of single women's land rights and the vulnerability of their livelihoods need to be understood historically, taking into account both the long-term changes in the colonial period and the immediate history of political turbulence, economic stagnation and AIDS in Zimbabwe. The conceptual basis of the argument outlined above is captured using the concepts of gender, land tenure, access, livelihood, vulnerability and pathways. The relevant concepts which are presented under the relevant subheadings reveal the multiple connections between phenomena that are the subject of inquiry. Using the concept of gender relates to the need to maintain different units of analysis. The livelihood framework is useful for looking at different kinds of resources along with the related concept of vulnerability. The concept of pathways analytically ties individual change over time to historical change and is useful for understanding individual decision-making in rapidly changing circumstances. The concepts also highlight the methodological challenges which had to be considered in the field.

There is a large body of literature on gendered land rights of rural women in Southern Africa, which suggests that the secondary land rights of these women leave single women vulnerable to poverty and destitution, particularly in a time of HIV/AIDS. This thesis suggests that the reality is more complex in rural Zimbabwe: firstly because gendered land rights are not well captured by the distinction between primary and secondary rights, and secondly because rural livelihoods are not just dependent on land. The main debates about gendered land access in patrilineal societies have primarily concerned the typology and the security of women's access relative to that of men. A major debate has revolved around whether claims to land are hierarchically ordered and gendered, with men having primary control and women having 'weaker' secondary rights (Mvududu and McFadden, 2001:110). Primary land rights give direct access to the resource and include rights to bequeath and dispose of land, whereas secondary rights are normally restricted to use rights. Other writers have rejected both the primary/secondary rights distinction and the hierarchical ordering of claims, stressing instead the existence of multiple claims, and the negotiated, dynamic and fluid nature of tenure relations (Gray and Kevane, 1999; Whitehead and Tsikata, 2003). The emerging pattern was that, although gender was a social construct and diverse, in patrilineal societies women generally had less control of resources than men. However, local level structures have demonstrated great diversity and complexity in women's land interests and factors affecting them (Razavi, 2003; Whitehead and Tsikata, 2003).

The communal farming areas of Zimbabwe are a specific type of legal and political space that was created by the colonial government to function as a labour reserve. Land access is the basis for communal area livelihood organisation. The internal governance of communal areas resembles other patrilineal indigenous tenure systems in the world. The structuring of gender relations in this space is specific and central to the understanding of land and natural resource access in communal areas. The Shona, for example, constitute a patrilineal society with an extended family system. Under this system, property and authority are vested in the male head (May, 1983; Folbre, 1988; Muzvidziwa, 2002). Patriarchy supports and justifies relations of subordination of women in their interactions with men (Muzvidziwa, 2002).

The concept of gender is important for analysing ideas about resource access and livelihood activities. Gender refers to the widely shared expectations and norms within a society about appropriate male and female behaviour, characteristics and roles (Gupta, 2000). Gendered patterns refer not only to relations between men and women as husbands and wives but also to relations between sisters and brothers, between co-wives, between mother and daughter, father and daughter and others (Peters, 1995). In communal areas of Zimbabwe, gender has generally been thought to determine women's access to resources as it defines their identity, position, entitlement and status through social-cultural meanings, practices and power (Kesby, 1999; Hindin, 2002). Traditionally, the institution of marriage has occupied a key position in a configuration of powers that constitute the space, identity and the dependent position of women in communal areas. The discourse and practice of mobility through marriage is important to a woman's identity (Kesby, 1999:30; Adams, 1991a). Married couples live in the husband's homestead among the Shona. Customary marriage laws and tradition give men control over women and their reproductive capacities (Folbre, 1988). Women's land rights are secondary to men's and women depend on the maintenance of a relationship with men to maintain access to key resources. Unmarried adult women are viewed as impermanent as they are destined to marry and move to another community (Kesby, 1999:30; Adams, 1991a). Consequently, the women are never allocated primary land rights (Gaidzanwa, 1994; Andersson, 1999). In the colonial era, women who were orphaned, quarrelled or had weak ties with men were vulnerable and tended to migrate out of the communal areas to the urban areas as they could not manage without agnatic or conjugal relationships to men (Gaidzanwa, 1994:12). However, some scholars have argued that the gendered hierarchy based on male dominance has been changing over time in Zimbabwe because of urbanisation, migration, and changing marital and livelihood forms (May, 1983; Folbre, 1988; Muzvidziwa, 2002; Vijfhuizen, 2002; Goebel, 2005).

Between 1890 and 1980, Zimbabwe was a British Colony. Colonialism initially destabilised pre-colonial gender relations through education, migration, urbanisation and religious conversion, which saw women exiting the patrilineal relationships (Cheater, 1986; Schmidt, 1992; Kesby, 1999). The colonial government and patrilineages co-operated to subdue women and reassert patrilineal control. Women were boxed into a status of permanent legal minority that increased their dependence on men

(Folbre, 1988; Schmidt, 1992; Barnes, 1997). After Zimbabwe gained independence from the colonial power, the increase in women's access to wages weakened gender hierarchies (Rwezaura *et al.*, 1995). Though post-independent Zimbabwe witnessed legal reforms that pushed through liberal feminist reforms, the changes were structured to exclude the communal areas. The result was that patriarchal discourse, practise and ideology reinforced male dominance in these areas (Gaidzanwa, 1994; Kesby, 1999). Customary tenure in communal farming areas fulfilled a specific function in the colonial administrative system as it reinforced patriarchy and male dominance and was therefore left in force by the post-independence government.

Male dominance in communal areas has traditionally been based mainly on the governance of agricultural resources, but the shifting livelihoods of communal area residents have witnessed a change in the role of agriculture. This change has altered gendered land access. In the context of this research, increased demand for land and the emergence of a land market pose serious threats to patriarchal control. Cash-based land transactions have undermined the powers of traditional authorities to allocate land to their heirs. Migration and social upheavals caused by land reform, economic decline and HIV/AIDS have undermined the marital relationship, which had been thought to be the dominant mode through which women access land (Scoones *et al.*, 1996; Huisman, 2005; Jackson, 2007). Recent work on widows' land rights has illustrated that widows may retain access and cultivation rights but mostly as guardians on behalf of their (male) children (Gray and Kevane, 1999; Jackson, 2003; Razavi, 2003). Single women's land rights may also be allocated outside the existing norms and a focus on practice is therefore warranted in the search to understand single women's resource access. Thus, the forms and significance of gendered land rights have changed over time in Zimbabwe. To understand questions of women's present-day land access they must be located analytically within changes in the governance of land and other resources and within changes in rural livelihoods for both women and men. These are discussed in the next two sections.

1.3 Understanding change in land tenure and resource governance

The land tenure status of single women in communal farming areas of Zimbabwe can only be understood within an analytical framework that encompasses the historical processes, which occurred in the country. This is because these processes shaped the institutions and governance systems in communal areas to produce the tenure status that communal area inhabitants and specifically single women have. In order to understand the tenure outcome, it is important to examine the various institutions and mechanisms through which people gain and maintain access to land and natural resources in communal areas. The concepts of legal pluralism, access and land markets are the most salient.

1.3.1 Communal farming areas and legal pluralism

Property relations and access to resources in present-day communal areas of Zimbabwe are governed by plural systems of laws and legal orders, practices and mechanisms (Cousins, 1993; O'Flaherty, 1998). The concept of legal pluralism is a relevant framework for analysing multiple legal regimes like communal areas. Legal pluralism is that state of affairs, in any social field, in which behaviour pursuant to more than one legal order occurs (Griffiths, 1986; Merry, 1988). Social fields include the family, the village, the community and the wider society. The communal areas of Zimbabwe are semi-autonomous legal fields in that there exist within them norms and sanctions with respect to land which are valid within a limited social frame and political and geographical space, and are accepted as legitimate by the interdependent actors there (Moore, 1986; Griffiths, 1986).

State law derives legitimacy from the State and is exercised through the legal system as represented by the judiciary and legislative systems. It is viewed as individualistic, rational, equally applied, written down, abstract, and has clear procedures (Griffiths 1986). State law has limited influence on land transactions in rural areas of Sub-Saharan Africa (Moore, 1986; WLSA, 1995; Andersson, 1999; Nyambara, 2001a; Whitehead and Tsikata, 2003).

Customary law consists of customs, traditions and institutions that directly influence people's lives and mediates access to land in villages, families and between individuals. Customary law is flexible and better able to respond to specific conditions. This provides opportunities for customary law's adaptation by individuals and individual families, and for regional variations. Customary law's flexibility and its ability to respond to specific conditions make it a more dynamic framework and may even increase the chances of traditional institutions ruling in support of women's claims against those of men (Whitehead and Tsikata, 2003).

The limit of formal law and legal pluralism in such a rapidly changing environment means that formal laws are frequently by-passed as people may take advantage of loopholes in the system. The laws and practices governing resource access in communal areas have increased over time with various reforms so that the relationships between various layers of legislation and land allocation practices have become unclear. The pressures on the communal areas resulting from various successive government interventions have also led to practices that are more responsive to the need of the populace than the official laws and rules. Such practices can only be discerned through a framework that goes beyond the legal-centred approaches to property relations and incorporates non-legal ways of accessing land. The concept of access (Ribot and Peluso 2003) provides a framework for analysing divergent means of exploiting resources, which is wider than the concept of property.

1.3.2 From property to access

Access can be defined as the 'ability to derive benefit from material objects such as persons, property, institutions, social and political, economic relations, actions, entitlement, relations of production and their respective histories that shape benefit flows. Different circumstances change the terms of access and may change the specific individuals or groups most able to benefit from a set of resources' (Ribot and Peluso, 2003:153). Access primarily refers to the social and political relations mediating access. Property refers to a more restricted set of relations that people have with land. The concept of property is normative and has spatial boundaries (Blomley, 2004). It is supported by regulatory systems of law, custom or convention (Ribot and Peluso, 2003). This view excludes the people who are not entitled to hold property (Rose, 1994; Ribot and Peluso, 2003). In communal areas, land cannot legally be annexed or transacted on the market and does not lend itself to regulation by state law as is the case in freehold property regimes. The property focus disregards derived land rights, which are most important for women as customary law prohibits them from holding primary land rights.

By using the concept of access, we can show how some groups or individuals control resources, whereas others maintain their access to resources through those who control them. Gaining access is the process by which access is established. In order to maintain access, people also expend resources. Access maintenance and control are complementary activities. Ribot and Peluso's concept also identifies different mechanisms of resource access, which are divided into rights-based and structural and relational mechanisms. The former consists of legal and illegal means that people use to access resources. Right-based access means that the State controls resources, and resource users have to invest in relations with the state agent to benefit from resources (Ribot and Peluso, 2003).

The ability to benefit from resources is also shaped by specific socio-economic, political and cultural factors within which access is sought. These include technology, capital, markets, knowledge, authority, identity and social relations (Ribot and Peluso, 2003). The access framework recognises that labour, social relations, knowledge, authority and identity are all possible mechanisms for gaining and maintaining resource access. Since women are often not formal right holders, a focus on these alternative mechanisms for accessing resources expands the scope of the research.

1.3.3 Land markets

Although it is illegal to sell communal area land in Zimbabwe (Pankhurst, 1991; Nyambara, 2001a), market transactions in communal area land frequently take place. The market transactions have been labelled informal or vernacular land markets (Andersson, 1999, 2001; Mbiba, 2001; Nyambara, 2001a, 2001b). These land markets operate informally under the customary tenure regime (Chimhowu and Woodhouse, 2006a). The land sales are disguised as sales and purchases of buildings and other improvements such as trees (World Bank, 1986 cited in Nyambara, 2001a:265).

The growth in communal area informal land markets is ascribed to population increase, land scarcity, mobility, inadequate urban housing and low wages (Andersson, 1999; Nyambara, 2001a; Peters, 2004; Chimhowu and Woodhouse, 2006a). Chimhowu and Woodhouse (2006a) have identified three categories of land buyers in communal areas. These are: 'new big men', who use income from a full-time job, and knowledge and influence from bureaucratic office to take advantage of opportunities in agriculture (Berry, 1993); migrants, who usually resort to vernacular markets because they lack customary rights (Woodhouse *et al.*, 2000 cited in Chimhowu and Woodhouse, 2006a; Nyambara, 2001b); and those with rights to land through kinship but, where land is scarce, have to resort to land purchase or rental often from a senior male relative with land to spare (Almanor and Diderutuah, 2001 cited in Chimhowu and Woodhouse, 2006a). The challenges of understanding informal markets arise from their locally specific dynamics of social and economic change and the difficulty of establishing what is being sold and bought, land improvements, use rights or community membership (Andersson, 1999; Chimhowu and Woodhouse, 2006a). This is important for understanding the insecurity experienced by single women who frequently change marital status, identity and residence because of disruptions in their lives. The socially embedded nature of vernacular land markets means that customary authorities are best placed to gain from commoditisation (Nyambara, 2001a; Woodhouse 2003 cited in Chimhowu and Woodhouse, 2006a), by personalising communal land that they hold in trust for the community.

Generally, the impact of informal land markets in communal areas ranges from increasing ownership by those with access to remittance income, distress sales by those affected by HIV/AIDS or the emergence of an informal market for re-allocation by tribal leaders of land owned by the deceased. (Chimhowu and Woodhouse, 2006a:360). The impact of the vernacular markets on women's land rights is diverse. In some cases, land markets have strengthened women's rights to control land and afforded the women protection from dispossession under customary law. The markets also discriminate against women however, as markets allocate land according to purchasing power (Mvududu and MacFadden, 2001; Nyambara, 2001b; Chimhowu and Woodhouse, 2006b).

1.4 Understanding changing livelihoods

In order to analyse how communal area residents make livelihood decisions, it is necessary to examine how they organise to gain and maintain access to resources, the challenges they face and the actions and decisions that the people undertake to confront the challenges to their livelihoods. In trying to capture the decisions that people make to confront challenges, it is necessary to use a concept that focuses on the individual decisions in light of the resources at their disposal as revealed by the lived experiences. The livelihood framework, vulnerability, agency and pathways concepts are used to facilitate an analysis of how communal area residents make livelihood decisions. These concepts are individually explored in this section.

1.4.1 The livelihood framework

A livelihood can be defined as 'the way people get by and get things done' (Bebbington, 1999:2021). The livelihood framework identifies five forms of capital at the basis of a livelihood: human, social, physical, natural and financial, which people use to build their livelihoods (Carney, 1998; Farrington *et al.*, 1999). The capitals allow survival and are the basis of power to act, reproduce, challenge or change the rules that govern the control, use and transformation of resources (Bebbington, 1993:3).

The institutions outside the domestic unit are important in determining access to resources and managing risk. The institutions may also provide safety nets to their members as a form of insurance against risks and threats (Scoones *et al.*, 1996:34; Huisman, 2005:260). These institutions are better analysed by focusing on social relations. Social relations are the key to understanding the capabilities of livelihoods. Using the social capital lens, we can explore the different ways in which social relationships can be mobilised from within civil society to manage resources of various types and engage other actors (Bebbington, 1999:2037) and gain access to social security. The social relations provide hearth-holds with direct and indirect support, physical, financial and moral support in times of need (Rwezaura 1989:5; Scoones *et al.*, 1996; Griffiths, 1988). This support is referred to as social security in this study. The institutions within which hearth-holds derive social security occur at the level of the family, village, wider formal and informal networks comprising friends and neighbours, the church and the State (Murray, 1980; Sharp and Spiegel, 1985; Guyer and Peters, 1987; Pankhurst, 1991; Dhemba *et al.*, 2002; Hartnack, 2005).

The family consists of an individual and his/her immediate kin and wider kindred or agnatic lineage (Murray, 1980). The extended family is the traditional social security system and its members are responsible for the care of the sick and vulnerable and the transmission of skills. In Zimbabwe, these include security, support and care of orphaned children (Foster *et al.*, 1997; May, 1987:81). The social security for widows in many African societies used to be assured by widow marriage to the deceased husband's brothers. The situation of orphans shows great divergence between custom and practice. The type of support depends on the life cycle stage of households and individuals within them (Griffiths, 1988). However, kinship-based social security should not be romanticised as its impact varies between particular settings (Murray, 1980:147; Schott, 1988; Golooba-Mutebi, 2005).

Traditionally, the village consists of a community whose members may be related through a common male ancestor (Murray, 1980). The traditional social security system allocates more land to the village head (*sabhuku*) under the *Zunde raMambo* (chief's granary)[3], because it is the *sabhuku*'s responsibility to host strangers. The *zunde* field is also used to grow food crops to feed the old, destitute and visitors in the village. The community provides free labour to work on the *zunde*. The control of *zunde* lies in the office of the *sabhuku* (Nyambara 2001a:262, Dhemba *et al.*, 2002:132).

Networks are loosely defined relationships outside those of immediate kin upon which an individual can rely (Murray, 1980:148). They include neighbours and the church. Scoones *et al.*'s (1996) study in a rural area in Zimbabwe noted that networks are important for accessing labour, loaning of draught animals, herding livestock and work parties. These arrangements enable single women hearth-holds to be cared for in illness, to mobilise household labour and to manage their survival.

Traditional religion acts also as a kind of insurance against insecurity. Ancestral spirits are consulted through mediums to mediate, arbitrate and advise, explain deaths, illness and provide protection. People continue to practise religious tradition because of fear and the need to maintain family ties. The latter is important since most people rely on the family for support when ill, unemployed or old (Vijfhuizen, 2002:223).

A study in rural South Africa showed that people prefer kin relationships for sustenance and support where these are available (Sharp and Spiegel, 1985:145). Women tend to concentrate in same sex groups, religious groups and wider family networks (Maluccio *et al.*, 2003:147).

In Botswana, the State has provided security through transfers of, for example, seeds and agricultural inputs in times of drought. This has enabled the poor to produce their own food. Public works jobs provide irregular but important source of income. The state intervention has, however, undermined co-operation and sharing arrangements (Solway, 1994). In Zimbabwe, many households in communal areas

[3] *Zunde raMambo* (chief's granary) is a pre-colonial traditional social security arrangement designed to address the contingency of drought or famine. The practise requires the chief to set aside land for the communal growing of crops for food security. The community assist with labour on a voluntary basis. The harvest is stored and distributed to the community when there are food shortages. The food from the *zunde* is also used to feed the chief's assistants, widows, orphans and the disabled. The programme was undermined by colonial government through systematic destruction of the traditional leadership's authority and the assumption of the food security guardian role by the State. In the mid-1990s, the Zimbabwe government tried to revive the concept of *zunde* in response to concerns about growing malnutrition and orphans. The success of the programme has been hampered by a shortage of land, inputs and labour (Dhemba *et al.*, 2002). In Gokwe, *Zunde raMambo* land was appropriated by individuals (Nyambara, 2001a).

came to depend on state transfers during the 1990s (Cousins *et al.*, 1992:5). The poor became more vulnerable when the State failed to provide the transfers due to budgetary constraints (Scoones *et al.*, 1996).

The social security system has been changing due to increasing mobility, individualisation of property, the weakening of traditional beliefs in sanctions from ancestors against those who neglect the needy, and feelings of abuse and exploitation among the different parties (Schott, 1988; Rwezaura, 1989; Bourque, 1997; Dhemba *et al.*, 2002). A study by Pankhurst (1991:612) in villages located in the same province as the present research area concluded that villages in the communal areas of Zimbabwe had changed from the core of men related through agnatic kinship with their wives and children to include far more unrelated people and many people who were considered members but did not live there. This induced changes in the social security system as the terms of benefits had to be revised to exclude or accommodate the diverse populations. People who relocate become isolated and lose contact with their kin (Sharp and Spiegel, 1985; Hartnack, 2005). The loss of access to reliable sources of cash income due to relocation, retrenchment and economic processes undermines networks of reciprocity and condemns most households to dropping out of reciprocity networks at times when those networks are needed most (Sharp and Spiegel, 1985; Hartnack, 2005; Scoones *et al.*, 1996).

A study of Chivi communal area in South Eastern Zimbabwe noted that successive droughts destroyed social security networks and wiped out cattle, which were used in draught sharing arrangements. This resulted in more individualised arrangements between those with draught animals and those without. In some cases, traditional forms of assistance have been replaced by monetised services (Scoones *et al.*, 1996; Huisman, 2005:260). *Zunde raMambo* is no longer being implemented in all communities because of problems such as lack of fertile land and inputs, and poor community mobilisation (Dhemba *et al.*, 2002). The traditional authorities, custodians of *Zunde raMambo* land, have begun selling *zunde* land on the informal market to obtain the proceeds for their own use.

The dominant position of elders has been undermined by the emergence of other forms of authority based on religion, politics through the entry of modern party politics and patrimonial rule through the ZANU party machine, and access to money. Elders are increasingly viewed as a liability and burden and increasingly vulnerable to neglect (Rwezaura, 1989:6).

These processes have also increased social and economic differentiation. Social and economic differentiation provides fertile grounds for suspicion and accusations of witchcraft, which erode social cohesion (Scoones *et al.*, 1996; Golooba-Mutebi, 2005). The church, HIV/AIDS and government laws have changed beliefs and practices around traditional religion. The old and young take advantage of the beliefs and strategically combine them to explain difficult circumstances and control others in resource struggles. The manipulation and use of beliefs has had direct consequences for women who are often accused of witchcraft and blamed for causing illness (Vijfhuizen, 2002:223; Andersson, 2002).

Families extend through marriage. The dissolution of, and failure to formalise marriage has depleted the extended family and reduced the opportunities through which single women negotiate access to resources (Foster *et al.*, 1997:156; Vijfhuizen, 2002:222). Consequently, some roles have broken down and modified, and others, such as widow inheritance, have disappeared. This has led to increased fostering[4] and child-headed households. Some of the reasons that people advance for refusing to assume responsibility for orphaned children are their illegitimate status, stigma and economic burden. The maternal grandmother remains an important source of support for orphans (Foster *et al.*, 1997: 163).

This review of the social security system has emphasised the importance of kinship, networks, co-operation and increasingly money and the State in mediating vulnerability. Social security is an important resource for individuals in need. Since there are various levels of institutions that potentially provide social security, it is not possible to make generalised conclusions about the impact on the various single women. The increasing numbers of single women, ill people, children and poor have increased demands on existing systems of social support.

With all these ongoing changes in the institutional framework of the communities the household's viability depends on its members' capacity to sustain or increase their access to livelihood resources, opportunities to turn those resources into sources of livelihood enhancement and means of enhancing existing ways in which those resources contribute to institutions and relationships (Ellis, 1998). Within a household, different members make contributions to the portfolio on the basis of their age, gender and endowments (Ellis, 1998).

'Livelihood diversification is the process by which rural families construct a diverse portfolio of activities and social support capabilities in their struggle for survival and in order to improve their standard of living' (Ellis, 1998:5). Diversification may occur as a deliberate strategy or an involuntary response to a crisis (Scoones, 1995; Ellis, 1998). In Zimbabwe, it has resulted in agricultural extensification, migration, rural industry, and rural and peri-urban commerce (Scoones *et al.*, 1996; Bebbington, 1999). The household's capitals determine the opportunities and constraints for diversification (Ellis, 1998; Francis, 1998; Bebbington, 1999). Diversification does not necessarily have positive, tangible or monetary outcomes (Farrington *et al.*, 1999).

[4] Mother's relatives fostering their grandchildren.

Nowadays it is rare to find people making a living exclusively from land in Southern Africa (Scoones *et al.*, 1996:3). Rural people combine farming with factory work, mining, farm labouring, trading, craftwork, beer brewing, manufacturing, work in shops, schools and local authority, prostitution or domestic employment (Scoones *et al.*, 1996; Francis, 2002). The livelihoods of the successful reflect flexibility and responsiveness to opportunities (Francis, 2002:539). The most vulnerable households are those that have no regular income and are unable to reproduce themselves without external assistance from other households or the State (Cousins, 1993; Francis, 2002). Many of these households have come into existence as a result of marital breakdown, the inability of young adults to establish independent homesteads, or mobility. People moving in from commercial farms lack a secure base from which to reconstruct their livelihoods. They lack networks and support and are excluded from resources (Sharp and Spiegel, 1985; Francis, 2002). The former commercial farm workers and single women are some of the most vulnerable groups in the communal areas of Zimbabwe.

The holistic nature of the livelihood framework enables an understanding of the diversity and multi-dimensionality of the poverty problem (Kaag *et al.*, 2002). The livelihood framework facilitates a focus on how livelihoods are constructed across the spatial divide and across the different arenas of resource access. It also facilitates an analysis of the economic, social and political relationships in a way that understands that these relationships are subject to negotiation (Bebbington, 1999:3; Murray, 2001). The approach views rural livelihood in monetary and non-monetary terms, both of which are complementary. By integrating rather than replacing other analytical approaches, the livelihood framework overcomes biases of disciplinary approaches and facilitates the analysis of a heterogeneous sample of people and their livelihoods (Farrington *et al.*, 1999).

Several critiques have been levelled at the livelihood framework. It has been criticised for understating the impact of wider processes, such as globalisation, which have a significant influence on livelihoods (Cahn, 2002; de Haan, 2000; de Haan and Zoomers, 2005). Globalisation through structural adjustment has undermined the national economy of Zimbabwe and resulted in the destabilisation of rural livelihoods. The resultant migration has weakened household structure and solidarity-based patterns of social security (Bryceson, 2000; Kaag *et al.*, 2002). This has increased the vulnerability of single women hearth-holds. The livelihood framework underestimates the importance of history and consequently fails to tackle the root causes of poverty (Murray, 2001; Francis, 2002; Yaro, 2006). The approach downplays the role of transforming structures, mediating processes, institutions and organisations – for example culture which is an important influence on the vulnerability concept (Cahn, 2002; de Haan and Zoomers, 2005). The role of institutional factors such as land tenure, village-based institutions for labour and wealth sharing, markets and state organisations (important factors for explaining differences in endowments and strategies) are considered only in passing by the livelihood framework (Van Dijk, 2002). The livelihood framework's neutral approach to power relations understates the role of imbalances in the perpetuation of poverty and inequality (Ashley and Carney, 1999; Cahn, 2002; de Haan and Zoomers, 2005). Households with inadequate responses to known livelihood risks are vulnerable (Chimhowu and Hulme, 2006:735). The livelihood framework does not adequately treat the vulnerability context and can potentially straitjacket investigations into the presumption that it is possible to stretch the capitals incrementally to enhance livelihoods (Chimhowu and Hulme 2006:747). In order to understand the vulnerability of single women, it is important to explore the concept of vulnerability in general.

1.4.2 Vulnerability

The concept of vulnerability is useful for unravelling how single women experience the threats to land access and livelihood maintenance in a patrilineal society and the uncertainty brought about by pressures on the communal land area resource, and how they manage threats to their human capital brought by mortality, morbidity and mobility. Vulnerability is defined as a high exposure to risk, shocks and stress and difficulty coping with them (Chambers, 1989). Shocks can destroy assets or force people to dispose of them in order to cope. A shock can be negative or positive (Scoones *et al.*, 1996). Shocks may undermine rural livelihoods in ways that increase vulnerability and result in impoverishment and powerlessness (Scoones *et al.*, 1996:9). The threats to communal area livelihoods arise from natural hazards (e.g. drought), disease, or human behaviour (opportunism, breakdown of sharing mechanisms, war and economic policy). Households face a variety of risks, and different people at different stages of their lives manage risks and uncertainty differently. Spreading risk across activities allows a diverse response to uncertain events and the contrasting dynamics of the operating environment. The ability to spread risk depends on access to, and control over, resources (Peters, 2004; Scoones *et al.*, 1996; Wisner *et al.*, 2004). The rate at which conditions change in communal areas is in itself a source of risk. The vulnerability framework allows the analysis of how women deal with those risks. The concept of vulnerability draws attention to the multiple dimensions of deprivation, such as gender, social exclusion, poverty dynamics and established coping patterns, and the resilience displayed by those affected (Yamin *et al.*, 2005).

Vulnerability is also a forward-looking concept that seeks to describe how prone individuals and families are to being unable to cope with uncertain adverse events that may happen to them. Vulnerability means not lack or want but exposure and defencelessness. It has two sides: the external side of exposure to shocks, stress and risk; and the internal side of defencelessness, meaning a lack of means to cope without damaging loss (Chambers, 1995:175). The concept seeks to describe the risks that people confront, the anticipatory management of those risks and what happens when those risks come to pass in the form of mitigation, coping and outcomes (Chambers,

1989). The broad factors causing vulnerability in Southern Africa are growth failures which have resulted in poverty and loss of migration options, market liberalisation failures, HIV/AIDS, politics and governance (Ellis, 2003). Vulnerability is embedded in complex relations of power, resource distribution, knowledge and technological development (de Haan, 2000; Eakin, 2005). Although the poor are relatively more vulnerable, they differ in their vulnerability as demonstrated by the differences in range of coping strategies (Watts and Bohle, 1993; Bohle *et al.*, 1994; Eakin, 2005), differing incomes, commodity bundles (Sen, 1981), relations with institutions and the decisions made by the affected populations (Eakin, 2005; Bohle *et al.*, 1994). Households headed by women are particularly vulnerable as they have no capital reserves such as money or other livestock, which they can liquidate to cushion their households (Bohle *et al.*, 1994).

1.4.3 Agency

The single women under study are neither powerless objects nor free agents. In order to understand how people deal with changing socio-economic circumstances and rural-urban linkages, it is important to focus on their ingenuity, and therefore agency in the diverse circumstances in which they find themselves. Bringing human agency to the centre of inquiry allows the researcher to analyse people's different responses to similar structural circumstances and at the same time to understand how individual choices are shaped by larger frames of action and meaning. Human agency refers to people's capacity to integrate experiences into their livelihood strategies and seek outlets for ambitions or solutions to problems. Human agency enables them to reshape social conditions Giddens 1987, de Haan, 2000). Agency is embodied in the individual but also embedded in social relations through whom it can become effective. Through human agency, structures may change. Human agency informs decisions and actions of people that cause livelihood changes in the end (Giddens 1987, de Haan, 2000). Women who challenge gendered stereotypes push boundaries and shift communities' expectations about gender roles and relations. This forces the community to rethink its long-held assumptions about power and may lead to a reconfiguration of local gender relations. This study, by focusing on 'abnormal' women (outside the marital relationship), uncovers evidence of shifting gender relations. For example, Schmidt (1992) details how women took advantage of economic opportunities and alternative judicial arenas as they sought sanctuary from traditional institutions in communal areas with missionaries and urban areas, as a result of which regulations were altered. The colonial state yielded to this pressure by introducing passes to allow some women to legally reside in urban areas and eventually provided housing for married couples (Barnes, 1997). Vijfhuizen's (2002:165) research conducted in a rural area of Zimbabwe after independence concluded that divorcees have to fight hard to maintain access to their ex-spouse's plot, including by going to court and other strategies.

1.4.4 Pathways

Agency most clearly manifests itself in the decisions that women taken to deal with risk in their environment and to ensure the viability of their livelihoods. The environment of communal areas in Zimbabwe is an outcome of long historical transformation through colonial and post colonial era. Since the environment in which single women in communal areas of Zimbabwe made livelihood decisions was characterised by rapid change over a number of years, it was important to identify a concept that reflects how individuals make decisions to cope with vulnerability in dynamic conditions. The concept of pathways highlights how people's strategies unfold as they interact with the environment and the fact that they are more a result of these interactions than of anything pre-planned (de Bruijn and van Dijk, 2005). Pathways arise out of iterative processes in which goals, preferences, resources and means are constantly reassessed in view of new unstable conditions (de Bruijn and van Dijk, 2005). Pathways analysis starts at the level of the individual. This facilitates the analysis of diversity in location, experience, assets, political position, social networks, social cultural variables, gender and generational differentiation. The concept of pathways is oriented towards an analysis of the dynamics of decision-making processes to pinpoint under what constraints and opportunities actors are likely to follow specific pathways to mitigate instability. The pathways concept is a relevant concept within which to study how single women make decisions in a fast-changing environment because its analysis is historical, it recognises that people's actions are not always consciously acted out, people dispose of diverse capitals to play the game, decisions are led by emotions and the framework recognises that decisions are taken in multiple environments (2005:11).

1.5 Understanding single women's livelihood decisions: units of analysis

1.5.1 Looking beyond the household to the hearth-hold and land units

The ways in which single women access resources and make livelihood decisions have to be assessed in terms of individual women, yet these individual decision-making units are also embedded within larger domestic groups. In the current research, this has been achieved through the selection of a representative kind of domestic unit, which required rethinking the concept of the household. The neo-classical household model is frequently used for analysing domestic units. This western concept, which assumes a male head, unitary decision making, and western gender relations based on a conjugal relationship, does not fit into the African residence, production, decision-making and consumption patterns (Peters, 1995; O'Laughlin, 1995; Ekejiuba, 1995; Francis, 2002; Quisumbing, 2003). The household is not a discrete entity but has permeable boundaries as it is embedded within wider structures, and it depends, especially in Africa, for basic livelihood and identity on encompassing structures (Guyer and Peters, 1987:205; Chant, 1997:6). The term 'head of house' is loaded with additional meanings that reflect the traditional emphasis on households as undifferentiated systems in a patriarchal system (Varley, 1996; Buvinic and Gupta, 1997). The focus on head of houseship renders invisible those single women maintaining families and residing in larger households, for example young single mothers residing in larger families and older unmarried women living in households conventionally regarded as male headed (Varley, 1996; Buvinic and Gupta, 1997; Chant, 1997:7).

The concept of female-headed households emerged in the course of a feminist critique of a non-gendered analysis and practise of economic development (Peters, 1995). A distinction was made between *de jure* and *de facto* female-headed households. The former represented women heading households in their own right, whereas the *de facto* female household heads were those heading households in the absence of men. This typology has also been criticised for homogenising female-headed households. Single women do not share common economic circumstances, and some female-headed households with economic means can be better off than those headed by a male (Peters, 1995; Izzard, 1985; Buvinic and Gupta, 1997). Varley (1996:505) argued that discussions of women-headed households are often based on the stereotyped image of a single mother with young children to the detriment of elderly widows and others heading a large proportion of women-headed households. The normative definitions of domestic units focus more on the absence of men than on the presence of women. This marginalised women who have never married, or are divorced or separated with or without children, elderly women living on their own or with an unmarried son. The limited utility of conventional patriarchal frameworks rendered the conventional units of analysis irrelevant and therefore there was a need to reconsider the domestic unit. The concepts of hearth-hold and homestead are more relevant for the study, which focuses on single women and their access to communal areas of Zimbabwe.

1.5.2 Hearth-hold

In a search for a more gender-sensitive analytical framework that recognised the importance of women, Ekejiuba (1995) introduced the hearth-hold, which refers to a female-directed social unit. It is structured on the mother-child bond. The hearth-hold is a unit of consumption and production. It is not necessarily a co-residential unit. It can be independent or interlinked in a household. It depends on transfers from other hearth-holds and households in the community. The circumstances that give rise to female-directed hearth-holds include polygyny[5], leviratic unions, three or more generational households of parents and their children, and brothers and their

[5] The state or practice of having two or more wives or concubines at the same time (Webster's Dictionary 1976 edition cited in May, 1983:51).

divorced/widowed sisters. The result is several hearth-holds nested in the household but not always headed by a male. The hearth-hold is a unit with more clearly defined and less contestable social relations. A hearth-hold is not destabilised by divorce, widowhood or non-marriage and thus it is a less volatile institution. Using a hearth-hold as a female-centred unit of analysis is a more direct way of reaching women. The hearth-hold concept facilitates a more effective analysis of gender and intra-household relationships. It enables us to see women as active, often independent, actors who shoulder responsibilities, take risks, strive to maximise livelihood options and positive impacts of their efforts on their dependents (1995:60). A focus on the hearth-hold as an independent unit of analysis facilitates the understanding of the significance of male and female adults as different agents of development and change (1995:56). The hearth-hold places single women in focus. This enables the identification of single women in their diverse locations as head, sub-unit, dependent or nested unit in more conventional households. Hearth-holds are more autonomous than the normative paradigms claim. They may have to conform, but hearth-hold heads remain in control of at least part of their individual circumstances. Hearth-holds have to mobilise external resources to survive. This organisation occurs across domestic and community boundaries. In order to take into account the various levels at which resources are mobilised, other units of analysis have been incorporated in this research, such as homesteads and the position of women within homesteads. This was done because, in the end, for anybody to be able to live in a communal farm area he/she needs a form of access to a place to reside, a homestead, whether legitimate or not and recognised by the community or not

1.5.3 Homestead

Female-headed households and hearth-holds were thought to represent transient and repeated phases of being without an adult resident man (Peters, 1995; Buvinic and Gupta, 1997; Izzard, 1985). The HIV/AIDS pandemic, the decline of polygamy and the reduced economic capacity of men have undermined the domestic unit and challenged the existing norms about adult women's living arrangements. It is problematic within the household to take into account 'members' who move in and out in response to the various stimuli. Since migration is one of the enduring responses of communal area residents, it is important to have a concept that allows the incorporation of migrated members into the domestic unit. The homestead is one such entity.

A homestead is the specific residential location to which male-headed households and female directed hearth-holds are attached. The homestead is a flexible patriarchal landholding grouping of kin who compose and recompose themselves into a number of constantly shifting residential patterns according to their changing and continuously negotiated interests. A homestead is a place where a family normally consists of a husband, wife and children, and men and women who originated from that homestead but live elsewhere. Children of deceased/divorced wives often stay on the homestead (Russell, 1993). Wives of sons can also reside on the homesteads with their children, as would divorced and unmarried daughters. Widows also form part of the homestead composition. Homesteads may provide residence to several hearth-holds belonging to the various women (Russell, 1993; Vijfhuizen, 2002). The composition of the group resident at any time is the outcome of the interplay of several, often divergent, interests of its members. Because the homestead is the collective property of the patrilineage, it precedes and succeeds members. Homestead membership is indeterminate but is always bigger than the resident group. Those residents at the homestead at any time are the members who happen to be exercising their right to live there. Absent members lose neither their homestead membership nor their kin-based responsibility to contribute to the support of various homestead members (Allen, 1990; Russell, 1993; Vijfhuizen, 2002).

Hearth-holds form part of the homestead grouping along with other male headed and conjugal domestic units. The advantage of the homestead over the household is that it encompasses a wider grouping of people. Unlike the household that focuses on a group of people, the homestead focuses on a socially defined space. This same space is a tangible unit used in the allocation of residential land rights in the study area through which people move. The homestead is a place from which migrants can move out but maintain a status and relationship with the communal areas. This elasticity allows the research to overcome physical limitations of tying individuals to specific places. This is an important consideration in studying the rights of women who move in search of better livelihoods. The homestead is also a space within which hearth-holds interact with other domestic units. These relationships are important for understanding hearth-hold differentiation and resource access. In addition, other land units represented by the arable land, gardens, pastures and other various land use categories will be used as units of analysis. This is to facilitate an analysis between the hearth-holds and the spatial dimensions of livelihoods.

1.5.4 Hearth-hold differentiation

Understanding the differentiation of hearth-holds is useful for explaining the differences in resource access and decision-making. The work of Chayanov on the Russian peasant household (Thorner *et al.*, 1966) is the basis for most household differentiation analysis. According to Chayanov, differences are a result of demographic differentiation, which correlates the size, and relative prosperity of households with their place in the cycle of generational production. Although the literature differs on the number of stages necessary, most models try to identify major transition points within the life cycle. These include marriage, childbearing, childrearing, a period of joint spouse survival without children and the eventual death of the spouse (Norton, 1983). Cain (1978) studied households in a patrilineal Bangladeshi society where private ownership of property and ownership of land was concentrated among men and inheritance law and customs favoured men. Cain's study established that the duration of a household's development cycle is determined by the length of a patriarch's tenure. The developmental cycle is inadequate for understanding household differentiation as it treats female-headed households as deviant (Murray, 1987; Francis, 2002). Murray's (1987) examination of materials on entire life courses of individuals provided insights into individual heterogeneity. In this way he demonstrated the development cycle model's limited capacity to deal with variation. The more relevant criteria for differentiating hearth-holds are identity, residence type, marital status, age, rank, economic status, health and the extent of kinship support.

Identity as determined by place of origin differentiates hearth-holds as it determines people's access to resources in communal areas. Immigrants have differential access to resources and the State to which the traditional authorities are the gatekeepers (Shipton and Goheen, 1992:310; Francis, 2002:547). This criterion makes immigrant women more vulnerable than autochthons in communal areas.

Hearth-holds in the communal area may be located in the marital residence to which they moved upon marriage or they may be residing at a place acquired through purchase or employment on the market. Other hearth-holds reside in their natal village, that is, the village in which the woman was born. Individuals turn to their natal family in times of crisis (May, 1987:37). The conditions under which single women access the various residences differ and are influential in single women's decision making. The various residential locations also determine the vulnerability of children belonging to these single women.

The life history of single women, whether divorced, widowed or never married, indicates the route by which women enter single woman status and also explains varied circumstances (Chant, 1997:11). For example, it is suggested that widows generally have the respect of rural communities in Zimbabwe, whereas divorcees are considered a social failure (Huisman, 2005:260). This differential perception of hearth-holds influences the opportunities available to single women in their communities. Widows are more likely to have acquired assets from marriage than divorcees and women who have never married. This makes widows relatively less vulnerable.

The age of a hearth-hold head may determine how long she may remain single. Life course is also important from the view of resources, in that households with older children may have fewer dependents as older children share in domestic burdens and are potential wage earners and providers of remittances (Murray, 1987; Pankhurst, 1991; Armstrong, 1995; Chant, 1997). Older children potentially cause fewer mobility constraints and provide opportunities for income-generating activities of women. Chant (1997:13) cautioned against the universal use of the life course as different contexts induce deviations in the normative course. The historical importance of labour for agriculture placed a high premium on large families and the reproduction of labour. As a result, the life cycle of a woman was centred on the reproductive and postmenopausal stage. Previously, women who became widowed while they were still in their childbearing stage would remarry and continue bearing children. Young widows were a rare occurrence, and the women who remarried continued the household development cycle as described by Cain (1978). The HIV/AIDS pandemic has destabilised this life path for women. Men are reluctant to marry HIV/AIDS widows because of fear of contracting the disease, and for the same reason widow inheritance is on the decline. This has made young widows a normal and increasingly accepted occurrence in rural areas of Zimbabwe (Huisman, 2005:260). This development has transformed the meaning of age and has to be incorporated into the differentiation analysis. It means that there are young widows, a previously uncommon category of women, who experience vulnerability in different ways to the elderly widows. Younger widows still in their reproductive stage are more likely to re-marry. This renders them vulnerable to the risk of contracting HIV/AIDS and giving birth to more children without spousal support. The opportunities and constraints experienced by women in the colonial and post-colonial periods differ considerably and should be taken into account when differences in single women's livelihood decision making are being analysed. As Chapter 3 will show, the colonial government placed mobility restrictions on women. The post-colonial government lifted restrictions on women's mobility. Chapter 7 will illustrate how these developments made women vulnerable in different ways.

Single women hearth-holds can also be differentiated according to their political position. This comes with age or local elected office and determines an individual's access to resources (Cheater, 1986; Rwezaura, 1989; Shipton and Goheen, 1992:310). A higher political position or connections with high-ranking men increase single women's control of resources.

Women gain economic status from having a regular source of income or from the inheritance of assets from a deceased spouse. The higher the socio-economic status of a single woman, the more capacity she has to raise children alone or live independently (Armstrong, 1995; Chant, 1997). Though single women are found in the lowest and highest classes, they tend to be over-represented among the poor in Southern Africa (O'Laughlin, 1998:10). The women with no wage link lack labour and struggle to produce their own subsistence. They rely heavily on transfers (Pankhurst, 1991; O'Laughlin, 1998). These women are more vulnerable. Pensions may provide security for women in their old age, but this is limited to a few women whose husbands were civil servants. Even for those women, security has been eroded by the high inflation that characterised the Zimbabwean economy during the research period.

HIV/AIDS has introduced another differentiating factor among hearth-holds by heightening the importance of health. Single women who are ill or hosting ill adults tend to be more vulnerable. HIV/AIDS constrains single women's access to resources by increasing dependency and reducing potential access to remittances. Ill single women are more vulnerable as this makes it more difficult for them to diversify their livelihood portfolios. However, ill, single women with access to money could overcome their vulnerability by negotiating resource access on the market.

The extent to which hearth-holds can ensure social security through kinship support contributes to differentiation among single women. This is discerned by living arrangements which range from single person, extended household, single-sex households, female-dominant households, grandmother-headed households and embedded female-headed households. The diverse living arrangements indicate the single women's positioning in relation to accessing resources like labour, social security, childcare, money or accommodation (Chant, 1997). These relationships reduce the vulnerabilities of some single women.

1.5.5 Extending the boundary of spatial inquiry: rural-urban connections

Although attempts to define domestic units centre on a specific physical space such as the homestead, village and communal area, the communal areas in Zimbabwe are linked with other spatial entities like urban, mining, commercial farming areas and growth points. As a result livelihood organisation transcends the spatial boundaries of the homestead. It is therefore necessary to extend the spatial boundaries of enquiry beyond communal areas to reflect the full range of networks, opportunities and constraints experienced by single women. The communal area, homestead and village remain the focal point, but their limitation is that they confine subjects of study to specific places. In order to inquire into the whole spatial domain of livelihood organisation in Zimbabwe, the thesis will use the term rural-urban connections (Andersson, 2001). Connections refer to the diverse movements and transfer of people, goods and services on the rural-urban continuum. Rural-urban linkages have been previously framed in terms of classic dichotomies with mutually exclusive rural and urban spaces. This dichotomy confined people to one or the other space and misrepresented the ties that bind the rural and urban sectors (Andersson 2001:83). Rural-urban connections describes the close ties that exist between rural and urban centres and acknowledges that the two spaces are not separable but rather exist as a single social universe. Focusing on rural-urban connections rather than migration allows us to study the diverse ways in which rural and urban areas relate and the different types, duration, volumes ebbs and flows of those movements of people and livelihood capitals on the rural-urban continuum in response to the various stimuli. While some people leave rural areas for economic reasons on migrancy labour contracts, others make opportunistic journeys for social or political reasons or all three. The movements on the rural-urban continuum are neither unidirectional nor systematic nor can they be generalised by gender. The movements cannot be explained purely in economic terms. The focus on rural-urban connections allows us to reflect this diversity. While state regulation and economic factors are important in the formation of rural urban connections, they do not determine them (ibid 2001). Individual actors interact with the economic and state factors to determine the nature of rural-urban connections at any point in time. Andersson's research found that the communal area provided a rural base in terms of social security for the travellers he focused on. The same research also concluded that rural urban connections were not just determined by economic interests but include opportunities to develop human capital through education, obligations, custom, and social ties that transcend the communal area.

1.6 Research aim and research questions

Single women are a fast-growing section of the population of communal farming areas. They fulfil a large number of productive and reproductive tasks, without recognition and support by policies. There is a dearth of knowledge on how single women access resources and take livelihood decisions in communal areas, because knowledge frameworks on land tenure and resource governance and policy have marginalised single women. Previous studies have shown that they are increasingly vulnerable (Pankhurst, 1991; Gaidzanwa, 1997; Huisman, 2005; Izumi, 2006). Some have suggested that the weakness of women's customary land rights is a source of this vulnerability, yet the meaning of custom in such a context of turbulent change is in itself contested. This research seeks to clarify these issues by answering the following central question:

In this period of heightened vulnerability, how do single women take livelihood decisions and gain access to productive resources in communal farming areas of Zimbabwe?

This central question can be broken down into a series of sub-questions addressed by the thesis:
1. How have historical changes in the wider environment of Zimbabwe affected resource access and livelihood decision-making in the communal areas of Zimbabwe?
2. What are the emergent modes of access to land and other resources through which women and men construct their livelihoods in the rapidly changing communal areas of Zimbabwe?
3. What are the specific vulnerabilities confronted by single women in obtaining access to land for residential and agricultural purposes in the communal areas in the context of these changes in the wider environment?

4. How do single women organise livelihoods and take livelihood decisions in the communal areas of Zimbabwe?

To answer these questions, fieldwork was conducted in two Shona villages in communal areas in Zimbabwe. The research focused simultaneously on in-depth studies of individuals and generalised study of the wider community in which the single women organised their livelihoods. Historic studies were used to take account of changes in the colonial period, the history of political turbulence, economic stagnation and HIV/AIDS in Zimbabwe.

This research contributes to existing knowledge on the communal areas of Zimbabwe. There is substantial research highlighting the impact of land reform on the former commercial farms (Kinsey, 1999; Chaumba *et al.*, 2003; Magaramombe, 2004; Goebel, 2005; Hartnack, 2005). The impact of the land reform, economic meltdown, political turmoil and HIV/AIDS on communal farming areas, however, has remained under-researched so far.

1.7 Structure of the thesis

The thesis consists of eight chapters. The first chapter comprises the introduction, the research problem and justification of the study. The chapter sets the scene for the rest of the thesis by de-constructing the dominant discourses around gendered resource access and livelihoods in communal areas. The chapter critiques the major theoretical approaches to conceptualising gendered resource access and livelihoods in communal areas, and identifies more relevant concepts for analysing single women's resource access and livelihood decision making.

Chapter 2 describes the research design strategy, research experience and the data collection methods employed. The research design is not a series of logical steps but an outcome of variances observed between the literature and practice in the field, the author's considerations in selecting the research area, villages and units of analysis. The chapter, through a description of the researcher's experiences, demonstrates the shortcomings of normative frameworks for investigating single women hearth-holds' decision making. The field experiences show that fast-changing environments require a sensitive and flexible approach.

Chapter 3 traces the history of gendered land rights and livelihood vulnerability in the communal areas of Zimbabwe, with emphasis on how single women came to be such an important social category. The chapter shows from a national point of view how an immensely complicated pile of institutions and laws was created to govern resource access in communal areas that have given rise to various institutional spaces within which single women negotiate resource access. The chapter describes how conditions in the communal areas have become increasingly unstable over the last ten years. This has increased uncertainty in the decision-making environment and induced livelihood change.

Chapter 4 describes the specific spatial and historical context of the study with particular attention to the ways in which two different communities and their leaders have responded to the recent crisis of land and livelihood in rural Zimbabwe

Chapter 5 explains the importance of rights to land for residence as an important aspect of the rural livelihoods of single women, showing that residence is not just space but recognition as a member of a community.

Chapter 6 explains the tragic puzzle of agricultural production in rural communal areas today – the presence of uncultivated land in the midst of scarcity – and shows how hearth-holds' access to agricultural land fits within it.

Chapter 7 shows, through a series of livelihood histories, the pathways followed by single women as they piece together different kinds of resources to confront (not always successfully) the insecurities of land and livelihood in rural Zimbabwe today.

Chapter 8, the conclusion, draws together all the discussions of the thesis to conclude, reflect on lessons learnt and illustrate the work's contribution to existing knowledge.

Chapter 2
Research methodology, design and data collection[6]

2.1 Introduction

This chapter describes the research area, the research environment and the methods used for data collection during the fieldwork period. It highlights the limits of standard assumptions and research methods for analysing communal area livelihoods. It also illustrates the theoretical, methodological and practical challenges of using conventional research methodologies in such a fast-changing environment and how over the research methodology developed.

The research fieldwork was conducted between August 2004 and December 2005. The timing was targeted to cover at least one calendar year to enable me to capture livelihood organisation throughout a calendar year. Field research was carried out in Makuku and Ndamba villages in Chikwaka communal lands, Juru Growth Point (see Section 3.2), Zimbabwe Widows and Orphans Trust (ZWOT)[7], and the High Court in Harare. The villages and the growth point are located in Mashonaland East province, a predominantly Shona-speaking part of the country.

As a Shona-speaking Zimbabwean woman, I shared a nationality and language with the community I studied. I was born and raised in a legal pluralistic society. I am married under both the customary and the civil laws of the country. However, my parents and my husband's family are predominantly urban based, though both families maintain land claims in the communal areas. In this way I lived and witnessed the changes in rural-urban connections which have such an important bearing on the lives of single women. The research problem is close to my heart as it is partly a reflection of my life and that of those around me. The extent to which the research inquired into my own life experience made it easier for me to engage the research community, as values and meanings overlapped to a larger extent than would have

[6] A version of this Chapter has been published Kakuru, D. and G. Paradza (2007) Reflections on the Use of the Life History Method in Researching Rural African Women: Field Experiences from Uganda and Zimbabwe. Gender and Development Journal Volume 15 (2): 287-297.
[7] The Zimbabwe Widows Association, formed in 1996, later to become the Zimbabwe Widows and Orphans Trust (ZWOT), is based in Harare. ZWOT was established to assist grief and poverty-stricken widows and orphans. ZWOT has in excess of 25,000 members, and new members continue to join daily. These members live in all provinces of Zimbabwe. This organisation is unique in Zimbabwe because it attempts to address the wide-ranging problems encountered by this marginalised population. The organisation assists with the legal, financial, medical and material concerns of its members. Additionally, it provides an important network of psychosocial supports for its members (Paradza, 2007).

been the case with a person of a different origin and language group. In carrying out the research, I was seeking to understand people's actions and the meanings behind those actions. It was inevitable that in some of the relationships and frequent interaction I became embedded in the families, homesteads and individuals with whom I interacted and this inevitably influenced the responses I received.

The research started with a literature and background study. This was important for framing the research. The findings of this phase are discussed in Chapter 3. The second phase was a reconnaissance visit to Zimbabwe to select the research sites and familiarise myself with the research context – a communal farming area in Zimbabwe in 2004-2005. The third stage consisted of the research design, which focused on the selection of study methods and data collection instruments. The fourth stage consisted of data collection. The final stage was the analysis and conclusions. The research design involved a lot of choices, all of which I cannot discuss in this thesis. Here I highlight the more significant choices and the post-fieldwork reflections.

2.2 The research area

The research area lies in the north-western part of Zimbabwe in Goromonzi district, Mashonaland East province. It is located about forty kilometres from the capital Harare. Map 2.1 shows the location of Chikwaka Communal area, the research area. Goromonzi comprises communal, resettlement and commercial farming areas, mines and small shopping centres. The research villages are located in a communal farming area. The choice of the research site was based on a number of considerations. It offered opportunities to explore the impact of the wider changes in the economy on the communal areas; the short distance to Harare, Zimbabwe's capital city, and the proximity of the growth point provided a better context within which to explore rural-urban connections. The proximity to a commercial farming area provided a unique setting for studying the impacts of the land redistribution programmes on the economic, demographic and spatial organisation of livelihoods in the communal area. The research site is also an area of high HIV/AIDS prevalence. This enabled me to investigate the consequences of the spread of the epidemic for land rights and single women. Lastly, I had previously worked in the area as a regional planner for the Government of Zimbabwe. This proved to be a useful negotiating point, as entry into communal areas was restricted because of the politicised environment.

Initially I planned to focus on two sites one nearer to and another further away from urban areas of Zimbabwe. I revised this after the reconnaissance visit. The widespread insecurity which characterised the fast track land reform and presidential elections in 2001-2 (McGregor, 2001; Kriger, 2005:2) made security a prime concern and I believed increased the risk of working in remote areas in the 2005 parliamentary elections. Zimbabwe was experiencing a fuel shortage at the time of the fieldwork and this also negated the selection of a more remote research site. I obtained formal

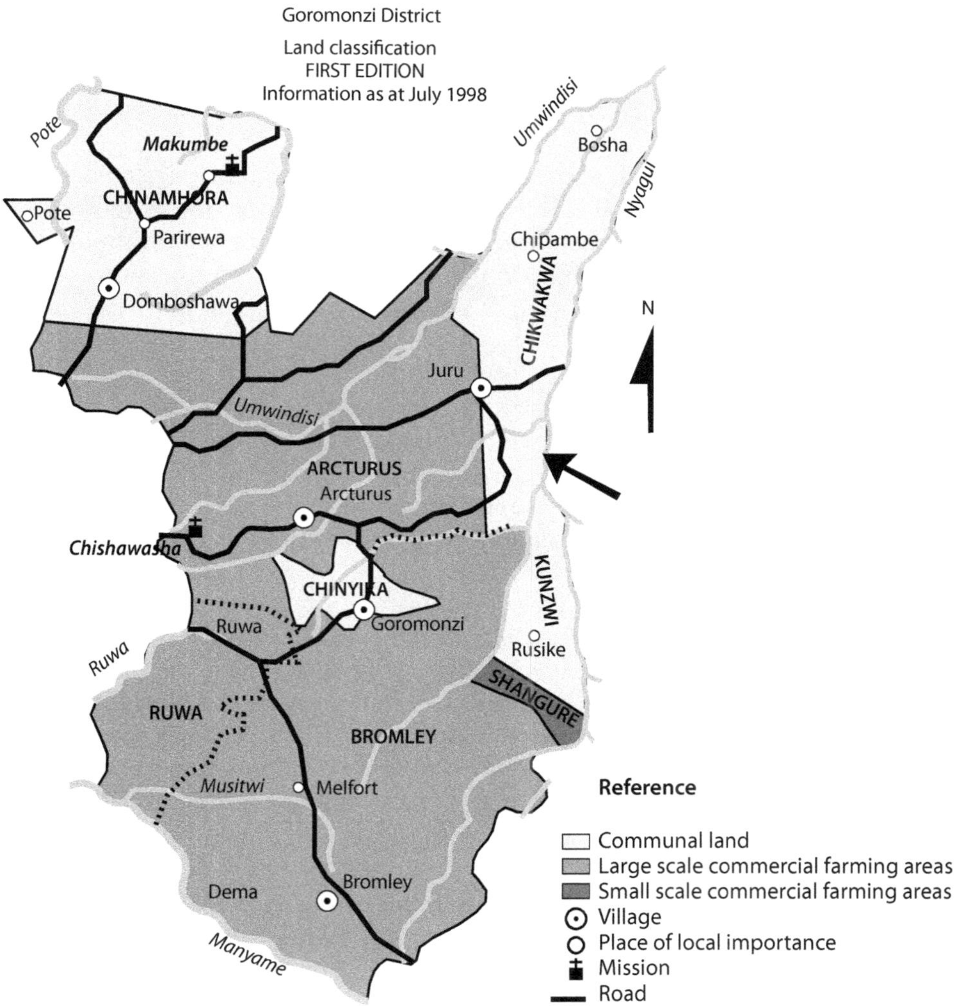

Map 2.1. Location of research area in Zimbabwe. Source: Surveyor General's office, 2008.

permission from various political, traditional and administrative authorities in the area.[8] This was a courtesy. The successful completion of fieldwork in an area characterised by political instability depends a lot on the co-operation of the various administrative and political institutions. Research conducted by the Women and Law

[8] I visited the various offices, introduced myself and the nature of my work. I also invited the various authorities to come and witness the research.

in Southern Africa Research and Education Trust (WLSA) noted that it was difficult to get information unless the conversation was prefaced by the researchers reassuring research subjects by telling them that 'We have been to the Chief' (WLSA, 1997:48). In my case, I referred to the clearance that I had obtained from the spirit medium.

The study largely focused on this one locality because of the context-specific nature of processes to be researched, i.e. livelihood decision making and vulnerability. The focus facilitated the contextualisation of the investigations, and this allowed a deeper understanding of the dynamics of political and socio-economic processes in the locality. These findings can be used as a basis for undertaking similar inquiries in other areas.

Goromonzi district is divided into wards. The work focused on Mwanza ward, which consists of 35 villages each of which is headed by a village head commonly referred to as a *sabhuku*. The village selection process involved my interaction with several people in the course of the two-month reconnaissance during which I attended gatherings, meetings, informal chats and village health workers' monthly meetings. I had intended to study a single village. As I made a shortlist, I realised that there were wide variations in village organisation, size, and population density, extent of land market and prevalence of hearth-holds. These variations and their obvious relevance for the study led to my decision to select two villages instead of one. The two selected villages represented extremes in size, organisation, resource access and leadership style, community activity, origin of inhabitants and degree of commercialisation of the land market. The differences in the two villages were significant in shaping hearth-hold vulnerability, livelihood decision making and people's access to resources, and are discussed in detail in Chapter 4.

Each of the two villages was subdivided into smaller, physically delineated land units known as homesteads. I made an inventory of all the homesteads in the villages with the help of four research assistants. The results of the survey were used to identify homesteads hosting single women hearth-holds. I used the shortlist of the selected homesteads to select case-studies for further inquiry. In selecting cases for focused research, I was not seeking statistical significance but to reflect the range of hearth-holds. I used age, marital status, village, number of residents, incidence of mortality and morbidity, socio-economic status and area of origin of homestead head in selecting the focus hearth-holds.

After realising that younger single-women hearth-holds were missing from my sample, I made further inquiries through which I learnt that the livelihood portfolios of single women were generally incompatible with full-time communal area residency. Single younger women were more mobile and relatively transient compared to the elderly counterparts. I interviewed younger women when they came to visit their parents in the study villages. I extended the spatial area of focus to include the growth point and Harare[9] where younger women were better represented. I focused on women who were members[10] of the two study villages to maintain continuity with these villages. Some embedded hearth-holds who lived as sub-units of the male-headed homesteads were also included. I selected 22 hearth-holds from 18 homesteads for in-depth studies.

I also targeted younger widows at the Zimbabwe Widows and Orphans Trust. For triangulation purposes, I held focus group discussions with key informants at the growth point. The 22 hearth-holds representing 18 homesteads in the study villages were the focus of the research.

The women who had moved away and/or were displaced from the study villages following a change in their circumstances were not available for interviewing. I replaced these women with the single women hearth-holds in the village that were made up of women who had relocated from other villages where they had migrated upon marriage or employment. I also used focus-group discussions and informal discussions with women (and sometimes men) that interacted with the departed women to get an insight into the displaced single women's decision making. These included neighbours, mothers-in-law and male relations. I derived several benefits from working with ZWOT. Firstly, it enabled me to locate the widows who left communal areas following the death of their spouse. Secondly, I got an insight into the work and impact of advocacy on property rights for widows. ZWOT provided a useful opportunity for me to interact with hearth-holds from other parts of the country and test the findings from the field site to ascertain their relevance to the rest of Zimbabwe. I attended meetings in Harare and in communal areas, conducted interviews, attended court sessions and interacted with ordinary members of the organisation over a period of six months. I held focus group discussions with ZWOT members during their monthly meetings and had individual interviews with widows.

2.3 Reconnaissance

The reconnaissance was a useful way of meeting people, discovering various issues that were of importance to the research area and testing the initial research framework. I randomly traversed the whole ward and used this opportunity to familiarise myself with the layout of the terrain, physical boundaries of villages and their social organisation as well as major physical features like roads, mountains and rivers. I capitalised on any gathering that was taking place in the ward to attend meetings. I used this opportunity to identify influential people in the ward and the topical issues at any point during the research period. I benefited from informal conversations with key informants about the situation in their respective villages

[9] Zimbabwe Widows and Orphans Trust offices in Harare.
[10] Membership here is defined as having access to residence in the village.

concerning population, mortality, livelihoods, gender and land transactions. I interacted with *masabhuku* by randomly visiting and spending time with them and enquiring about their area. This was a good way of getting to know the layout of the ward and relative location of places in the ward. I revised the framework by focusing on how domestic units worked and the specific ways in which people organised access to livelihood capitals. As a result, the conventional data collection approaches became inadequate and were revised or sometimes ditched. The next sections details how this happened.

2.4 Units of analysis

The units of analysis are the single women hearth-hold, homestead, land units (represented by homesteads, arable land, forest, garden and water channels), the village and ward. The use of multiple units of analysis is necessary because of the importance of linkages between the hearth-holds and other socio-economic spaces. The focus on hearth-holds as a unit of analysis represents a major departure from the conventional domestic units used in patrilineal land research. I explored beyond the individual hearth-hold to fully investigate the various social organisations, networks and places within which hearth-holds experienced livelihood vulnerability. The homestead is the physical location of the hearth-holds constituted around single women. The homestead encompasses single women living as daughters, widows, sisters or in other relationships. It is on the homestead that single women base their hearth-holds under differing residential and economic arrangements. These include living alone, residing with parents, siblings, children, employees or other arrangements. The homestead serves as a focus unit and a physical place where individual hearth-holds are based and organise their livelihoods. The homestead is also the administrative unit used by the State and conventional governance systems. In this study, the homestead was the unit at which the availability of assets such as the house, agricultural equipment, wells and toilet could be assessed. The incidence of hearth-holds on the homesteads was indicated by their presence and registration as homestead members during the inventory. These were either a head of a homestead or a co-resident of a homestead. The presence of foster children whose mother was divorced, widowed or deserted was also taken as an indicator of hearth-holds' access to the communal area homestead. This category included women who largely lived and worked elsewhere, women who left the village after the death of their spouse and those who had died and had left children in the homesteads.

The physical layout of the two villages consisted of individually demarcated homestead sites ranging in size from 2.308 to 0.492 hectares in Ndamba, and from 2.1745 to 0.130 hectares in Makuku. Homesteads were separated from each other by a fence or vegetation. Some homesteads had another piece of land for agricultural use located further away from the homestead, known as the arable land.

The village is an organisation with definite spatial physical location and spatial boundaries and has a legal recognition from the State. The research at village level enabled me to capture resources and processes that could not be observed at the homestead level. Each village consists of a collection of homesteads. In Mwanza, the number of homesteads in a village ranged between 20 and 300. The variation was due to the extent to which individual villages participated in the informal land markets, the governance style of the *sabhuku* and the size of the village. Although some villages like Makuku accommodated the market led demand for land, others like Ndamba resisted pressure and remained relatively small in number of homesteads. Intra-village dynamics were widely divergent. Makuku was one of the oldest villages dating back to the 1960s. Makuku village was selected because it emerged as one of the largest three villages in the ward in terms of population numbers and density. The village had 68 homesteads in 2004. The number of homesteads in the village grew to 90 at the end of 2005. In terms of surface area, Ndamba village was one of the smallest and 'youngest' villages in the ward. It was established in the early 1970s by young men who were living in Gosha village. Twelve families came together and requested the Chief for their own land to set up a village. This was granted and Ndamba was set up. The original Ndamba *sabhuku* was deceased and the current one was his son. The village had 23 homesteads at the time of the field research. The number of Ndamba homesteads remained the same throughout the research period.

2.5 Data collection

I used both quantitative and qualitative methods including interviews, a household inventory, mapping and observation. My goal was to obtain a deep understanding of the processes around communal land livelihood and the position of women within them. Here I found qualitative methods particularly useful. Although quantitative research provided figures that could frequently be used to say something about the position of women, the fact that positions, authority, gender identities were continuously changing made them more a matter for qualitative analysis (Vijfhuizen, 2002:233). Qualitative methods helped me to gain an understanding of the culture and community and provide a way to validate quantitative results (Quisumbing, 2003:9).

2.5.1 Inventory at village level

In order to contextualise the position of single women, it was important to have a general picture of the research setting. The aim of the village level inventory was to get an understanding of homestead membership, its size, the incidence of mortality, land access mechanisms, area of origin, assets, livelihoods and other basic data regarding the homestead. I also used the inventory to identify single women hearth-holds. The findings of the inventory are presented in Chapter 4. I administered a questionnaire that was both structured and open-ended to all homesteads that were occupied[11] between September and November 2004 in the two selected villages. I compiled a list of all individuals residing on the homestead, their relationship to the head and their demographic characteristics. In order to triangulate my findings, I used a list of village residents obtained from the *sabhuku* for the inventories.

This exploratory inventory is conventionally achieved through the surveys carried out at the initial stages of fieldwork. The rapidly changing population composition in the study villages rendered this strategy inadequate. Trying to record the actual number of people in a village was challenging because for a variety of reasons some remain unknown and unregistered in official circles[12] (Andersson, 1999; Nyambara, 2001a; Chimhowu, 2002:560). The inventory process had to continue throughout the field research period because of the rapid change in people's circumstances and high mobility of the population. The point of reference used in Chapter 4 is the situation at the beginning of the fieldwork.

The inventory proceeded relatively smoothly in Ndamba because the *sabhuku*'s list corresponded with the names of people encountered in the homesteads of his village. Fifteen homesteads were occupied during the time of inventory. Eight homesteads were vacant. The total number of homesteads in Ndamba remained at 23 after the GPS survey at the end of the fieldwork in 2005. The household survey presented in Chapter 4 focuses on the 15 occupied homesteads.

The Makuku inventory was more challenging. The reasons emerge in Chapters 4, 5, 6 and 7 and are important in understanding demographic dynamics of communal areas and their impacts on hearth-hold livelihood decision making. Initially I obtained an indication of the total number of homesteads from the *sabhuku* and village health worker to use as a guide. However, the two figures varied widely. I obtained a 'register' from the *sabhuku* who had a very short list of 25 homesteads. This was not a formal government register, but a book in which the *sabhuku* recorded names of the village members. After a day in the field and a high level of inconsistency between official names and those obtained from the field, the *sabhuku* supplied another register that was supposed to cover everybody in Makuku. This new list supposedly included Makuku 2 and Makuku 3. The new list had a total of 52 homesteads. I resolved to compile my own register as I proceeded with the inventory and try to reconcile the names and numbers later or at least, as was the final case, explain the anomalies. At the end of the Makuku inventory in September 2004, there were 68 occupied homesteads. There were 90 homesteads in Makuku at the end of 2005.

[11] Some homesteads were vacant because of death, or temporary or permanent migration.

[12] Andersson's study in Buhera communal lands described a similar experience where during a dispute over grazing lands a *sabhuku* who was asked to reveal names of people living on a grazing area deliberately concealed some names (Andersson, 1999:569).

There were names in the *sabhuku's* book that had not been accounted for. I had recorded names that did not appear in the *sabhuku's* book and the numbers could not be reconciled. There were tenants, purchasers, descendants and employees in the village, most of whom were highly mobile. This made it impossible for me to compile an 'accurate' comprehensive register of all residents by any criteria. There were a number of people who concealed their identity because they had purchased land on the informal market. Others provided the name of the original owner, tenants sometimes gave theirs or the name of the landlords, and children provided yet another name. This experience illustrates the challenges of data collection methods in anthropological research. Although this may be a problem for quantitative analysis purposes, it enriched my understanding of customary resource governance and illustrated an underreported but inherent characteristic of communal area livelihoods, that is, the high mobility of the population. The experience illustrates how the conditions of instability in rural Zimbabwe have induced rapid changes in the survey homestead membership and their individual circumstances. This is what led to my decision to focus on individual pathways in my presentation of the context of vulnerability in Chapter 7.

2.5.2 In-depth case-studies

The main data collection method used was the case study (Yin, 2003). Case studies are useful for answering the 'How' and 'Why' questions about a contemporary phenomenon within a real life context. The case-study observation probes deeply to analyse intensively the multifarious phenomena that constitute the life cycle of the unit with a view to establishing generalisations about the wider population to which that unit belongs. This was useful for this research, which focuses on the hearth-hold, homestead and village. The units for the case-study were selected so as to reflect the variety in life-situations and living arrangements of single women. The study went right down to the scale of the individual. In this way, I was able to elicit the experiences of the smallest unit in the field (Yin, 2003).

The case-studies, which focus on the 22 hearth-holds that I identified during the inventory, are located in the village, the growth point and in urban areas. The focus on the various spatial entities enabled me to study the individual hearth-holds' rural-urban connections. Chapters 5, 6 and 7 explain how hearth-hold decision making influences the location of hearth-holds. The research focus on the communal area villages has biased the sample towards hearth-holds that maintain a base in the villages.

In-depth studies were conducted of the selected homesteads and hearth-holds between August 2004 and December 2005. I undertook weekly visits to the homesteads to collect the hearth-holds' life histories and monitor their livelihood activities and decision-making. I employed a combination of instruments to collect information. I observed and informally interviewed other people who interacted with my selected hearth-holds. These included children, siblings, in-laws, relatives and friends. The focus of the research on single women in the study sites attracted the attention of other villagers who volunteered information for the in-depth case studies. Both positive and negative information was volunteered, but these voluntary participants added to the richness of the data and should not always be overlooked. One of the reasons why the research targets led to volunteers divulging negative information was because the research subjects (single women) have a socially stigmatised social status (Kakuru and Paradza, 2007).

Life histories generate reflections on topics that would otherwise remain implicit because they are taken for granted (Gysels *et al.*, 2002). Life stories reveal the complex, contradictory and multiple roles that women juggle. Life history methods have been found valuable for generating new insights into women's experiences not generated through other research techniques such as questionnaires and focus group discussions, since women's experiences are often ignored or silenced when they are at variance with those of men (Anderson and Jack, 1991; Francis, 1992). Life history interviews help to suggest connections and illuminate the causes and meanings of relevant events, experiences and conditions (Runyan, 1982). The usefulness of the life history approach depends on the researcher having a thorough understanding of macro-developments, which frame people's lives, as this is important for interpreting some of the responses informants give for their actions (Francis, 1992:92).

Life stories are often used to triangulate information from other data sources. Life stories themselves can be triangulated through interaction with people other than the research participant. Such sources in this research were gossip and other people who co-resided with the narrators, such as children. This led to inconsistencies in the form of different versions of aspects of an individual's story as told by neighbours, children or other family members. The inconsistencies posed methodological dilemmas but some information ironed out inconsistencies, and others were resolved over time (Kakuru and Paradza, 2007).

In life history research, trust is an important pre-condition for successful data collection. Trust was established through constant interaction and confidentiality. As trust and rapport were established over the course of the year, the quality of data improved, and the women divulged intimate details of their lives that would not have been recorded in the survey or focus group settings.

The targets for these life histories were the 22 hearth-holds selected for the in-depth case-study sample. The life histories were organised according to major historical and biographical events in the life of the respondents and were intended to facilitate a longitudinal insight into the events that shaped the women's status, the major decisions taken and the outcomes. Generalised probe lists proved to be of limited value because the circumstances that structured the experiences, memories and interpretation of individuals who participated in the study differed widely. Individual life events such as marriage, childbirth and death, which undermined or strengthened the welfare of their hearth-holds, had more biographical relevance (Kakuru and Paradza, 2007).

2.5.3 Key informant interviews

The key informant interview technique is pre-eminently suited to the gathering of the kinds of qualitative and descriptive data that are difficult or time consuming to unearth. Although emphasis is on qualitative data, key informants can generate some quantitative data. Key informants allow a researcher to develop a definition of dimensions involved in the research, to discover boundaries of communities, to identify extremes and to increase knowledge of the problem (Tremblay, 1957). The technique facilitates the collection of valuable information from knowledgeable members of society. These people also represent dominant discourses on how things should be run. Their views may not necessarily reflect the lived reality of all the people in the village. The key informants interviewed for this study included official or already recognised 'knowers' such as government officials, headmen, chiefs, lineage elders, local government ministry representatives, mediators, court officials, councillor, ZANU-PF representatives, the Master of the High Court, shop owners, ZWOT, agricultural extension workers and the police. Key informant interviews generated information on historical and factual data on land tenure evolution in the area, identified the main triggers of livelihood evolution and provided longitudinal data on how women's land access had evolved and the implications of this for future generations and rural livelihoods in general. Key informants provided community level information on their area of competence, such as migration, attitudes, beliefs, natural resource information, incidence of disputes and local demographic profile.

A weakness was that most key informants were male. Males are the gatekeepers of society in the study communities. They also had administrative and political authority to set parameters of community life (Vijfhuizen, 2002; WLSA, 1997:48). Therefore, women with specialised knowledge or with positions of community significance were specifically targeted to counterbalance male knowledge paradigms. Female key informants provided information on how women dealt with the formal structures of society (WLSA, 1997:49). In the research site the women with specialised knowledge included the elderly women, women who held political posts in the ward, village health workers and spouses of the village leadership, the spirit medium and commercial sex workers. Children proved to be rich key informants as they shared insights with me from their viewpoint. Children were important in revealing information on movements and life experiences of the single women. It was often through children that I learnt about the informal liaisons that single women had with men and the involvement of single women in illegal activities. The children's narratives were triangulated with information from other key informants and the women concerned.

2.5.4 Focus group discussions

Focus groups give an insight into a group's perceptions, attitudes, experiences and assumptions on a defined topic. Participants build on each other's ideas to provide a view that is not possible to capture at the individual level. The method also allows the generation of unexpected comments whose discussion may provide new perspectives. Focus group discussions are good for generating qualitative data and are an effective way to obtain general views and source issues in debates around the research topic (WLSA, 2002). They involve interviewing a group of people with specific expertise on a topic to discuss. The focus group discussions in this research centred on vacant land, tenure evolution, widowhood and property inheritance, land markets, the ways in which single women accessed productive resources in the communal area, livelihoods and governance. The method also investigated areas of consensus and diversity in the participants. Focus group sessions were conducted with *masabhuku*, men in Ndamba village, the Chief and sub-chiefs after the weekly courts, village health workers in Mwanza ward, single women hearth-holds who were based at Juru Growth point, widows from ZWOT in Harare, and members of a garden co-operative in Ndamba village.

2.5.5 Observation

Observations provided me with insights into the covert processes by which hearth-holds negotiated and maintained access to productive resources. I attended gatherings and meetings to experience and observe the interactions among the diverse groups. The public events demonstrated the interactions of the villagers and their leaders and gave me an insight into power relations and resources struggles. These included monthly *sabhuku* meetings, monthly ward meetings, village health worker meetings, funerals, field days, grain distribution, AIDS Day commemoration, Chief's weekly court, informal public discussions before and after formal public meetings, widows filing their property inheritance papers at the high court, Independence Day, tuberculosis (TB) awareness campaigns, election of National AIDS Council and ZWOT monthly meetings.

I participated by assisting with registration, balloting, distributing benefits and transport. I observed interactions between the hearth-holds, immigrants, indigenous occupants, various civic and political leaders. These interactions illustrated the tensions with regard to the identities, power relations and processes that any of the research subjects had been unable to articulate. For instance the formal meetings gave clues about what was happening behind the scenes. During a meeting to elect office bearers for an HIV/AIDS programme, I was appointed as the polling officer.

I listened as various eligible candidates were considered unsuitable for the various offices on the basis of their 'wearing the wrong t-shirt'[13]. This included an orphaned girl child who was nominated to represent orphans and vulnerable children. In another case, I turned up for meetings of a specific ward committee only to have the meeting postponed each time because the supposed chair and his deputy never attended the meeting. It turned out that by not turning up, the two gentlemen wanted to avoid the elections to elect new office bearers as this threatened their interests. In the time leading up to the National Presidential Elections in 2005, all public meetings were punctuated with political slogans. The locals were reminded by the leaders that all people in the ward were Zimbabweans and should support the Ruling Party. A month after the elections discussions about evicting 'immigrants and squatters' dominated the ward meetings. Lastly, throughout the fieldwork period, *sabhukus* were frequently asked by the government to compile and submit lists of vulnerable households. After each of these announcements, I regularly gave a lift to hearth-holds and former farm workers who wanted to go to the specific government office to verify that their name had been included on the list. Discussions with the people revealed that they did not trust the *sabhuku* to include their names on the list because of a variety of reasons, which included a failure and/or refusal to bribe him or a long-standing disagreement. I also regularly came across hearth-holds in my sample who made a point of attending every single meeting including those to which they had not been invited. The reason they gave me was that, they were scared they would miss out on some important meeting or they did not trust the village representatives to disseminate the correct information after the meetings. I also observed that there were some hearth-holds that never attended any public meeting during the fieldwork period. I probed this behaviour with both the hearth-hold and the authority that had called the meetings to gain further insight into relations and processes in the study villages. There were also tensions between the traditional, administrative and political authorities who struggled to exert control over governance in the communal areas. These are highlighted in Chapter 4.

Often I became an observer of daily life during life story interviews and when I went to offer condolences to homesteads in the study villages that had lost a family, member. This occurred when the narrative was interrupted by visitors or a narrator could not keep an appointment because of an unplanned family meeting, funeral or another gathering. During the 'observation phase' I stopped the interview but remained in the house while my in-depth interview subject interacted with her visitors. Though I had suspended the verbal interrogation, I listened to the discussions and used the materials from such sessions to complement the information I gathered through other interactions. In my observer position I was able to explore other facets of the

[13] Opposition party t-shirt.

narrator's life and community roles that would not have been exposed through the life history interview setting.

2.6 Land units

In order for people to live in communal areas, they need access to a land unit, as historically this has been the basis of communal area livelihoods and membership of a community. People use their land access for residence or generating livelihood through agriculture, leasing or selling, but land is the basis upon which they negotiate access to other resources and livelihood opportunities. Land units in the research area were represented by homesteads, arable land, forest, garden and water channels. The land units were a subject of inquiry and therefore formed units of analysis. I used a variety of methods to capture the different perspectives and means of accessing land. I walked through the villages with key informants who explained the lay of the land and land uses to me. I used the land register which I compiled after the inconsistencies experienced during the inventory in Section 2.5.1. I also compiled a vacant land register of and observed activities on vacant land, conducted focus group discussions, key informant interviews and mapping with the help of GPS to study land units.

There appeared to be a lot of unused land in an area of increasing land pressure. I focused on this anomaly and revised the research design. I compiled a vacant land register in June 2005 and during the GPS mapping exercise in November 2005. Over the research period, I enquired more deeply into these vacant land pieces in the two study villages through interviews and observation. The enquiry into vacant land culminated in focus group discussions during which a diversity of interest groups discussed the phenomenon of vacant land (*matongo*). These were *masabhuku*, widows at ZWOT and women members of the garden co-operative in Ndamba. The findings of these enquiries are discussed in Chapter 6.

2.6.1 GPS mapping

I used maps to get the ward layout of villages, types of land use and the spatial distribution of land according to its uses and to some extent the users. In Zimbabwe, there is a dearth of maps for use in analysing communal area land distribution. The maps that exist were produced by government to delineate communal farming areas from the commercial farms, urban areas and business centres. Layout maps for communal areas are rare. I set out to record land mappings of the research area. The maps normally used for this are those produced by the surveyor general's office on a scale of 1:50 000. Government maps tend to favour dominant (primary) land use and categories at the expense of minor uses and secondary land rights. They also portray one-dimensional notions of property with exclusive control and ownership that have obscured women's lands, resources, products and activities nested within men's property in rural landscapes. A multi-method approach resulting in pictures and words allows research to elicit explanations embedded in gender and life history (Rocheleau, 1995:463).

I duplicated a hand-drawn map in use in the clinic. The environmental health technician allowed me to copy the map which he had drawn up for mapping communicable disease prevalence in the ward. This map proved to be a useful base map as it showed an up-to-date presentation of the main roads, rivers, bridges and, more importantly, village boundaries.

I carried out a GPS mapping exercise at the end of the fieldwork programme in November 2005. The aim was to make land maps of the research villages because the situation on the ground differed from the official maps, which are out-dated. Other objectives were to quantify people's land holding, to create a database, to draw a village layout and to assess settlement patterns, to quantify the amount of land used versus under-utilised land, to capture land boundaries in a customary tenure area and to make visual records for the research. I was also interested in producing a visual representation of the division of land in the village by gender, generation and origin of the settlers. I enlisted the services of a surveyor and two assistants. I used the following equipment: GPS receiver, type: Trimble Pro XRS, GPS data logger, type: TSC1 Asset Surveyor, Pathfinder version 2.51.

2.6.2 GPS data collection procedure

The surveying team was taken on a pre-visit. This was important, as it enabled the community to get to know the surveyors and also facilitated the drawing up of the data dictionary. The pre-visit involved a physical inspection of the two villages to take note of the important features to be included on the maps. The next stage was the mission planning that involved devising a data dictionary in the office. We created the data dictionary using the Pathfinder software. The data dictionary is a file that is created in Pathfinder and contains the features (rivers, trees, fields, homestead, vacant, origin), and type of attributes and attribute types (polygons, lines, points, etc.) that will be best given the spatial information of the features to be surveyed in the field. The data dictionary was uploaded to a file in the GPS data logger (TSC1 Asset Surveyor).

Before the field visit, a few things had to be checked, including the GPS batteries, the cables and the settings on the receiver. Settings on the receiver included the coordinates system to be used in the field (for both Ndamba and Makuku, Universal Transverse Mercator Projection was used, Datum was World Geodetic System of 1984 [WGS84], Zone was 36 degrees South) and the logging options in the field, i.e. the frequency of data collection; in this exercise we collected data at 10 minute intervals. We collected data in the form of waypoints (single point data), polygons, and lines as defined in the data dictionary. All the data collected were corrected by the Activated OmniSTAR Satellite Differential Service, in other words the positions obtained by the receiver were corrected to fit the international standards by getting corrections via a satellite from internationally recognised trigonometric beacons in South Africa in real time.

We mapped the village boundaries for the two villages first. These are shown on Maps 4.1 and 4.2. The boundaries were meant to coincide with rivers, trees, valleys or other distinctive physical features. In practice, the situation was quite different as explained in Chapter 4.

The homestead borders are marked by boundaries and fencing. The mapping of arable fields was more challenging as here boundaries are more lax. The *sabhuku's* wife accompanied the team in Ndamba. The Makuku *sabhuku* initially accompanied the team but he handed over to some children after we had covered Makuku 2 and Makuku 3. The *sabhuku* accompanied us to allay his and the immigrants' apprehensions about the research and mapping exercise. The arable lands in Ndamba were lying fallow due to inadequate inputs and security problems. Some resembled bush. The guide skilfully identified individual arable field boundaries in the bush. These included trees, footpaths or contour ridges. The fields were distinguished by ownership and generation. The field data were downloaded using the Pathfinder software. Processing of the data involved creating Shape files and sending them to Arcview for the drawing of the map.

The GPS mapping exercise enabled me to re-visit the question of land access, and in this way, I obtained relatively reliable information about the vernacular land markets. The GPS mapping highlighted the insecurity of the communal area population. Individuals expressed fears that the maps would be used by the State to possibly withdraw under-utilised arable land, to evict illegally settled people and to detect and penalise those who extended their land boundaries beyond what was 'officially' allocated. I had the advantage of dealing with a relatively small area, which made it easier for me to recall and crosscheck responses. The focus on the actual activities, which took place on the land units in spite of their designated use, also exposed the relationships between different users and facilitated a multi-dimensional view of land access modes. Examples included people grazing livestock on a vacant homestead, homestead construction on the water channel, fencing of arable land by private individuals. There was an area in Ndamba, which for most of the fieldwork and in the writing is referred to as community forest (This is indicated in Map 4.2). This was because of the observed activities and type of enquiry that focused on the current use of the resource. During the GPS mapping exercise, the community forest emerged as a collection of individual arable fields, which had reverted to bush as a result of non-agricultural use. I had observed collective exploitation of the bush for pasture and firewood and assumed that this was an exclusively communal resource.

2.7 Reflections on the research design

Initially, the data collection procedures had been formulated and designed on the basis of normative assumptions about gendered land relations and livelihoods in communal areas of Zimbabwe, based on the traditional literature that paints rural livelihoods in a static way by presenting customs and traditions as static and denying women agency (Odgaard and Bentzon, 2007:202; see also Section 1.4.4). For many reasons, life in the communal areas has changed significantly. These reflections on the research reveal the adjustments that I made to the research framework to enable me to tackle the research questions. The following sections show how I adjusted the initial research framework in order to confront the practical, theoretical and methodological challenges that arose. The investigation of land access modes and the organisation of domestic units posed the greatest challenges. The initial research problem had been formulated on the basis of the hypothesis that agricultural land was a central livelihood asset in communal areas. In that regard, HIV/AIDS has undermined women's access to land and in so doing decimated women's food security. In arriving at that hypothesis, I had used standard assumptions linking land access to gender and food security.

I realised that my research framework was inadequate when I arrived in the communal areas and found land lying unused and single women maintaining access to land. As I proceeded with the research, it became increasingly clear that land access modes were not one-dimensional, and I revised the research design accordingly. I decided to focus on access instead of property and ownership, on hearth-holds and homesteads instead of the household (marital unit) and on pathways and vulnerability in addition to livelihoods. I sidelined the legal centrist line of enquiry and widened my focus to include all the possible means that people used to gain access to resources. This approach enabled me to capture the wide diversity in the communal areas resource access. I used single women and women key informants and diverted from the normative key informants of the traditional leadership and male key informants. I embraced the views of children and paid attention to the unsolicited responses that I received from the community. I viewed agriculture as one of the livelihood activities and did not focus on it as a central activity. I realised that the primary/secondary, rural/urban dichotomies and related tenure terminology was inadequate for framing single women's resource access. I went beyond the conceptualisation of rural and urban areas as mutually exclusive spaces and the division of men into worker peasants and women into farmer housewives[14] (Potts 2000a:808) to explore wider rural urban connections which cut across geographical and gender boundaries to explore the intimate connections between the urban and rural areas in Zimbabwe. I paid particular attention to women who had relocated following the end of their

[14] Worker peasants are urban migrant males who will eventually return to the land and farmer housewives are rural based women left to farm the land (Potts: 2000a 809).

marriage because I wanted to get a full account of their decision making. This was achieved through pathway analysis and analysis of life histories of the women in the study villages who had returned to, left and remained in the village after the end of their marriages. I recorded all the diverse ways in which people, and single women in particular, negotiated access to land and productive resources. My aim was to understand the decision making of single women over their whole life course and diverse locations.

I was preoccupied with security throughout the research fieldwork period. The study of land issues in Zimbabwe is highly politicised. Although I tried to remain apolitical in various capacities as a neutral observer, I was inevitably drawn into situations that compromised this. I frequently provided transport to people who would stop me along the way so that they could make announcements about political meetings and gatherings. I constantly worried about the impact of these activities and roles on the research community's perception of me and my study. I also made a decision to retreat from the research area for two weeks around the election period. This was because of the heightening tensions between rival political parties. In my absence, village residents who included a teacher and some school leavers were employed to record events. This not only facilitated a triangulation of sources, but also allowed me to experience events through the eyes of the key informants.

The respondents were uncomfortable with me using electronic equipment to record the interviews so I had to rely on handwritten notes. Initially it was difficult to conduct private interviews as people always came to sit in and listen. I was careful to stick to the non-political script. I attended as many meetings as possible and also provided transport to the leadership before and after meetings; this gave me a chance to follow-up on issues raised and observed in the field.

The study population was highly mobile as marital status and livelihood activities changed constantly. This posed challenges to the research. Although initially I had thought that the study of 22 cases was too ambitious, the high mobility, mortality and absenteeism rendered this a manageable sample. Hearth-holds' rural-urban connections were constantly reconfigured as they moved rapidly between the growth point, urban areas and communal areas. Living arrangements also changed rapidly in the case-study homesteads.

The research area has a high HIV/AIDS prevalence. HIV/AIDS is a highly sensitive and stigmatised subject. Some of the homesteads in the study villages experienced death during the research. It was not easy to ask direct questions about the illness. I used proxies of opportunistic illnesses[15] for HIV/AIDS. These included tuberculosis,

[15] illnesses that thrive on immune systems weakened by chronic infections (UNDP, 2004a:46).

pneumonia and meningitis. The community members were, however, surprisingly forthcoming in their responses, particularly when reporting the mortality of their own homestead members. Nevertheless, only one of the in-depth study women communicated her HIV status to the researcher. The rest did not know their status. People's willingness to talk openly about sensitive issues can be attributed to the fact that researchers are outsiders, the effect of HIV and AIDS awareness campaigns and anticipation of assistance from the researchers by the research participants (Kakuru and Paradza, 2007:290). I adjusted the research schedule in many instances to offer condolences, attend funerals that occurred during the research and pay constant visits to ill research community members, even though they were not part of the sub-sample selected for in-depth study. I maintained a respectable distance during mourning periods by not actively interviewing the bereaved but by retreating into observation. I lost three hearth-holds which I had selected for the sub-sample. Two of the hearth-holds relocated and one hearth-hold succumbed to HIV/AIDS. They were not replaced, but the information they provided up to their death and the fate of the survivors and resources were included in the analysis as they represented the single women hearth-holds that succumbed to the difficult living conditions in the communal areas. I also incorporated one hearth-hold that was formed during the research period. I encountered this hearth-hold when I enquired about a single man-headed homestead. By then the widower had taken a wife. Since the marriage was less than three months old, I decided to investigate further. I established that, in fact, the new wife had previously constituted a single woman-headed hearth-hold. As will be explained in Chapter 5, the man died and the woman resumed her single woman-headed hearth-hold status. The developments in this particular case revealed an enormous amount about the dynamics of single women's decision making.

Land access modes were challenging to investigate. This was because of the complex governance systems, limited records and reluctance on the part of residents to confirm market transactions in communal land (Nyambara, 2001a:266; Chimhowu and Woodhouse, 2006a:365). The residents who had sold and purchased land on the illegal market were reluctant to divulge the sources of land. Instead, the purchasers claimed to have obtained land from the village head as gifts, allocation under customary law or gifts from 'friends'. I got around the problem through a combination of inventory, life story and mapping. The relatively long research period entailed more interaction with the research community during which I was able to iron out most inconsistencies. The inventory captured data on area of origin of homestead occupants, reason for locating in the study village, father's occupation and length of stay in the village. This information yielded indicators of land access means used by the various homesteads. My privileged position as a native Shona speaker enabled me to use people's dialect as a criterion for judging whether they were original inhabitants of the area or immigrants.

In Makuku, the *sabhuku* who had initially allowed the research was reluctant to allow the team to proceed with the GPS mapping at the end. This was because of his own role in the informal land market and his discomfort with my increasing pre-occupation with the subject. It was around this time that the tensions between the immigrants and the autochthons in Makuku were high. The mapping finally proceeded after the intervention of the spirit medium and sub-chief. However, by this time it had rained and people were preparing the land for the next agricultural season. This delay cost me the opportunity to capture the differences between used and unused arable land from the previous cropping season in the village.

The exclusive focus on single women could have led to my arriving at a false conclusion that the vulnerability of single women hearth-holds was necessarily greater than that of other households. This was a possibility, but I could not have shown it to be true without looking at other kinds of domestic units. I took some steps to remedy this by including:

- an adult man living alone on a homestead;
- an elderly couple that had relocated to the village from a former commercial farm;
- a recently married young couple who were in the process of establishing an independent homestead;
- a widower who had married a serial single mother, Alice, who I incorporated in the case-studies.

This rapid change in the situations of people in the research area mirrors the rapid pace of change in the research community. The single women hearth-holds selected resided in a variety of residential arrangements, some of which included co-residence with men and women in a variety of economic relationships. My repeated visits to the residential homesteads enabled me to also interact with these other women and men who resided with single women and better appreciate their experience, although this is not stated explicitly in this thesis. The purpose of widening the in-depth study sample was to ascertain the extent to which the vulnerabilities of single women were common to the whole population. I drew from the experiences of these other people in my analysis in Chapters 4, 5, 6 and 7.

2.8 Conclusion

The research findings presented in this thesis are specific to the study area. The area's location in close proximity to commercial farms and the capital city provide a peculiar context that is not reflected in all the communal areas of Zimbabwe. However, the historical context that has shaped the vulnerabilities of single women in Goromonzi district extends across communal areas in Zimbabwe more generally. The findings thus provide useful insights into the processes occurring in communal areas in Zimbabwe in general. The findings of this work can be used as a basis for undertaking similar inquiries in other areas. Triangulation of data was achieved using a variety of methods and cross-checking. This enhances the validity of the research findings as each method supplemented and checked the others.

In this chapter, the methodological design experiences and data collection procedures of the research were reviewed. The limitations of conventional research methods and strategies used to address them were identified. The chapter has shown the volatile conditions under which the research was undertaken which necessitated short-term decision making by the researcher. The volatility arose from the political tensions around the election period, high mortality of the study population and rapid mobility and resident turnover in the study homesteads. The conditions necessitated fast thinking and continuous revision of the research design in the field at various crucial moments which are highlighted in this chapter. In adapting the research methodology as I proceeded with the fieldwork, I developed my own research pathway.

Chapter 3
Changing land governance and livelihoods in Zimbabwe's communal farming areas

3.1 Introduction

This chapter, based on secondary data, explores the historical and contextual framework of land tenure and livelihood decision making of single women hearth-holds in communal farming areas. The complexity of single women hearth-holds' land rights and the vulnerability of their livelihoods need to be understood historically, taking account both of long-term transformations in the colonial period and of the immediate history of political turbulence, economic stagnation and AIDS in Zimbabwe. The transformation of institutions in communal areas governing resource access has always been contingent upon larger economic and political developments in the country. This has created livelihood constraints and opportunities for single women and the communal area population at large. The legal complexity introduced by successive governments has created a diversity of institutional spaces, which have shaped productive resource access for single women hearth-holds and other people in the communal areas. The aim of this chapter is to discuss the context in which single women emerged and to provide a backdrop within which research results can be presented and analysed in subsequent chapters. The historical analysis of the evolution of single women's vulnerability and decision making in communal areas of Zimbabwe is a complex endeavour. The combination of factors makes this a complicated chapter.

In order to simplify it somewhat, the chapter is subdivided in five sections. The first, a largely descriptive section, presents an overview of the history of Zimbabwe's communal areas. The section describes how the communal areas were established and identifies the roots of vulnerability faced by communal area inhabitants. The second section highlights how developments in the last decade have exacerbated the vulnerability of the inhabitants of communal areas. It shows how the changing rural-urban connections increased the vulnerability of the communities in communal areas. The third section focuses on how a complex, ambiguous but male-dominated governance system developed in communal areas as a result of both colonial and post-colonial government policies. The fourth section focuses on how communal area residents deal with risks and access productive resources under such dynamic conditions. The final section concludes with implications of all the developments on single women hearth-holds' productive resource access and livelihood in the communal farming areas of Zimbabwe. The aim is to construct a framework within which to analyse how single women take livelihood decisions and gain access to productive resources in the communal areas.

3.2 History and rationale for the establishment and maintenance of communal farming areas in Zimbabwe

In 2003, 70% of Zimbabwe's population was living in rural areas (UNDP, 2004a:15). Communal area livelihoods in Zimbabwe have historically depended on a combination of farming and non-agricultural employment (Berkvens, 1997; Scoones *et al.*, 1996:67; Chimhowu, 2002:573) and rural-urban connections. The main crops grown are hybrid maize (the national staple), cotton, tobacco, millet, sorghum, vegetables, sunflower and groundnuts. Communal area farmers earn cash through sales of fresh produce and forest produce. The forests provide resources that the poor households depend on during droughts and periods of unemployment (Pankhurst, 1991; Cousins, 1993; Berkvens, 1997; Nyambara, 2001a). Livestock is the main store of capital for the communal area households. Farmers also own some agricultural equipment such as scotch carts, ploughs and wheelbarrows for agricultural production. Cattle ownership is biased in favour of those with access to a cash income. These include male-headed and *de facto* women-headed households (Pankhurst, 1991; Berkvens, 1997). Ox-drawn ploughs are the main source of draught for tillage. A diversity of rural-urban connections are a fundamental resource to the construction of livelihoods. Income earned in urban areas has been historically used to fund agricultural activities in communal areas and investment in capital development (Potts and Mutambirwa, 1990). However, connections extend beyond economic reasons to include a person's rights and obligations in the community (Andersson, 2001).

The relative contribution of agriculture to livelihoods in the communal areas of Zimbabwe varies from year to year, season to season and between different people depending on their gender, asset base or age (Jackson and Collier, 1991; Scoones *et al.*, 1996). During the 1992 and 2002 droughts, communal area maize output fell by more than 50% (Mudimu, 2003:23). The households relying solely on agriculture are the most vulnerable (Andersson, 1999; Coudere and Marijse, 1991: 70; Scoones *et al.*, 1996; Chimhowu, 2002:573). This is because of the variable climate and the incidence of drought. Investment in non-agricultural activities is necessary to reduce exposure to agricultural failure. Single women are more vulnerable as they lack agricultural inputs, labour and financial capital for agricultural production (Pankhurst, 1991; Berkvens, 1997). The people who can diversify from agriculture have more secure livelihoods. The communal area inhabitants engage in a variety of non-farm activities to diversify their livelihoods. These include wage labour, remittances, trading and off-farm labour. Single women hearth-holds remain marginalised in non-agricultural activities because of their reproductive and domestic responsibilities, limited access to capital and networks, and historical discrimination. The developments leading to the vulnerability of single women hearth-holds are better understood through a historical analysis of the establishment and evolution of communal areas.

Zimbabwe, the country formerly known first as Southern Rhodesia and then Rhodesia after Zambian independence in 1964, was a British Colony until 1980. The country had a population of 11.6 million in 2002. Of these, 70% resided in rural areas and were directly dependent on the land for their survival (UNDP, 2004a:11). Agricultural land in Zimbabwe is best evaluated in terms of its agricultural potential, according to which the land is divided into five regions: region I is for specialised and diversified farming. The region receives more than 1,050 mm of rainfall annually, evenly distributed. This is confined to the Eastern Highlands Districts and is well suited to tea, coffee and intensive livestock production. Region II, located in Mashonaland highveld, is an intensive farming region that receives 700-1,500 mm rain per year in summer. The area is suited to growing maize, tobacco, cotton, wheat, other grains and intensive livestock production. There is an associated sub-region sometimes referred to as Region IIb in which the research area is located. This differs from Region II in that the sub-region experiences greater rainfall variability and risk. Region III receives 500-700 mm rain per year. The area experiences seasonal drought and is best suited to semi-intensive crop and livestock production. Cropping is risky particularly for maize, which requires large quantities of moisture at specific periods. Region IV is a semi-extensive farming region, which receives 450-600 mm of rain per year and experiences frequent seasonal drought. This region is suitable for livestock farming and drought resistant crops. Region V is an extensive farming region, which receives less than 500 mm annual rain. Livestock farming is the suitable activity (adapted from Rukuni and Eicher, 1987). Communal farming areas are mainly in Regions IIb to V. These areas have fragile soils and are vulnerable to periodic drought. The agro-ecological conditions impose severe limitations on the ability of communal areas to produce food. Only nine percent of communal areas receive adequate rainfall for the regular production of maize. Six of the country's eight provincial districts in which communal areas are located have been consistently unable to meet local food requirements over the past decades (Page and Page, 1991:3; Scoones *et al.*, 1996; Mudimu, 2003).

There are 172 communal farming areas in Zimbabwe (Coudere and Marijse, 1991:72).[16] They were specifically created and maintained as a vulnerable space by the colonial and post-colonial government. The country was divided into land classes during the colonial era. The 1930 Land Apportionment Act resulted in the appropriation of fertile agricultural land in Regions I-III for European farming. The same Act confined Africans to infertile land on overcrowded Native Reserves (Floyd, 1962). The Act also demarcated economic space by designating land as urban, rural, mining and national conservation park areas. Rural land was further divided into commercial

[16] Pre-colonial Zimbabwe land tenure systems are a subject of divergent debate (Holleman, 1969; Moore, 1998; Cheater, 1990; Nyambara, 2001b) Various authors give different interpretations and explanations of the governance and land tenure situation in pre-colonial times. This makes it difficult for one to make a conclusive judgement about the pre-colonial land tenure systems.

and communal farming areas and small scale African Purchase Areas. Black Africans were allowed to hold land in private ownership in the African Purchase Areas. Urban, mining and commercial farming areas were for Europeans. Africans were not allowed to own land in European areas but could work there. The communal farming areas were created by the government to serve as an administrative jurisdiction, a labour reserve and a consumer product market for the commercial and urban sectors. Communal areas also served to keep Africans off European land and deny Africans urban citizenship (Duggan, 1980; Ranger, 1985; Amin, 1992).

The colonial government created communal areas as labour reserves. This was achieved by restricting Africans to infertile and overcrowded native reserves. The overcrowded conditions on the reserves restricted the Africans' ability to grow their own food and forced men to seek wage employment. This ensured the continuous flow of cheap labour for European enterprise in the form of cyclical migration between the communal farming areas and the commercial farms, urban areas and mines (Floyd, 1962; Page and Page, 1991).

The colonial government used a variety of strategies to undermine the viability of communal areas and entrench the communal areas' dependence on the urban areas and commercial farming and mining areas. The aim was to protect European farmers from competition and ensure a market for the produce from the European commercial farms. This was achieved through licensing and strict regulation of traders in the reserves (Cheater and Gaidzanwa, 1996; Pederson, 1997:72). There was under-provision of trading centres in communal areas (Wekwete, 1991; Cheater and Gaidzanwa, 1996; Scoones *et al.*, 1996). There was biased provision of research, advice, and credit and transport infrastructure to the European commercial farmers. The colonial government also used discriminatory pricing, imposed market levies and maintained unfavourable conditions of access to state marketing institutions for African maize growers (Duggan, 1980; Ranger, 1982, 1985; Ladley and Lan, 1985; Bratton, 1987; Page and Page, 1991; Amin, 1992). As a result, the communal areas became dependent on formal centrally controlled markets for selling their labour and produce, buying food and consumer goods, staple maize, agricultural inputs, cash, employment and services.

By denying Africans urban citizenship, the colonial government set up the communal areas as a social safety net for urban migrants. This was achieved through legislation forbidding African ownership of land in European areas. This 'tradition', started during the colonial era when the authorities decided that it was in the interest of tradition for every male native to have a home in the reserves, has been accepted by Africans to the extent that those who could have opted out of the reserves feel obliged to maintain their land rights. As a result, the dominant form of rural-urban connections was cyclical migration labour. This consisted of men relocating during their working life to work in urban areas and remitting their wages to the communal areas. The remittances were used to buy agricultural inputs, seeds, fertilisers, pesticides, cattle, draft and development of the homesteads. The women largely remained in rural areas and invested the urban remittances in farming and running of the homestead. The urban wages subsidized the rural area by providing financial capital while the communal area subsidise the urban by bearing the costs of raising and feeding the family and supplementing the urban wages with agricultural produce. The investments and obligations that these men who migrated out of the communal areas made helped them to secure and maintain their communal area membership and access to a place to retire to at the end of their work life and eventually a place to be buried on their eventual death. Possibly, immigrants' decisions to maintain their land rights may provide opportunities for marginalised groups to access land in communal farming areas.

The colonial government imported labour from the neighbouring countries of Malawi (formerly Nyasaland), Mozambique and Zambia (formerly Northern Rhodesia) to supplement the local labour from the reserves and keep wages low. These migrant workers were accommodated on the commercial farms and mines where they worked (Chadya and Mayavo, 2002; Magaramombe, 2004). Many farm workers were badly treated by their employers, lived under poor and unhealthy condition, and were paid paltry wages (Rutherford, 1997). The foreign farm workers were officially prohibited from getting land in the reserves and many returned to their home countries after finishing work on the farms (Rutherford, 2001). However, there were many who remained in Zimbabwe.

After the native reserves were established, the population in the reserves continued to grow and land pressure intensified. The colonial government feared that the rising human and livestock population in the reserves would undermine the government policy of land segregation (Drinkwater, 1989). In an effort to redress the situation, the Rhodesian government passed the 1952 Land Husbandry Act. The act aimed at facilitating colonial control of land use and land allocation in the reserves. Each African man was allocated a standard holding of arable and grazing land whose size was set by the State. The standard was six to eight acres depending on the climatic conditions (Andersson, 1999:561; Nyambara, 2001a:256). However, progressive farmers were allowed to purchase rights from fellow peasants. Polygamous men received extra land for each wife after the first, and chiefs and headmen received extra allocation in recognition of their duties[17]. Although rights were basically restricted to adult men, women who were divorced, widowed, over 25 and unmarried, or whose husbands were missing were eligible for a permit to hold land (Machingaidze, 1991:567; Phimister, 1993:227; Andersson, 1999:561). Although the Native Land Husbandry Act (NLHA) discriminated against married women by registering land

[17] To feed visitors, the old and destitute in the community (Nyambara 2001a:262).

holdings in the names of their husbands, the same Act recognised and acknowledged the needs of single women.

The NLHA increased differentiation in landholdings among the peasants, increased landlessness among Africans, dispossessed the men who were away working in European areas and increased the flow of indigenous Africans to urban areas (Machingaidze, 1991; Phimister, 1993; Ranger, 1993; Shutt, 2002). The NLHA implementation was halted in 1962 after covering 42-60% of the reserves.[18] The government identified the slow speed of the team implementing the programme, shortage of staff, doubts on the suitability of the Land Husbandry Programme for lowveld and increased land pressure as factors which made it impossible to meet the targets and an economic recession that curtailed the demand for wage workers. The land allocation patterns designed during the NLHA period remain the basis on which land management is practised in many communal areas of Zimbabwe (Duggan, 1980; Drinkwater, 1989; Machingaidze, 1991; Ranger, 1993; Andersson, 1999; Mbiba, 2001).

Between 1969 and 1979, Rhodesia was ravaged by a civil war which was mainly fought in the communal areas. The war disrupted livelihoods in these areas by disturbing production and inducing displacement (Lan, 1985; Gugler, 1989; Potts, 1995; Kesby, 1996). The war destroyed 180 rural clinics and 23 hospitals (UNICEF/Goz, 1985 cited in Loewenson, 1991:366). The disruption of veterinary services led to huge livestock losses (Lawrence *et al.*, 1980). Some of the communal area inhabitants escaped to the urban areas, whereas others moved to more fertile communal areas. In 1975, the urban population in Zimbabwe doubled (Drakakis-Smith, 1984:1284). The increase in urban population was accompanied by an increase in informal activities and urban agricultural activities, which were dominated by women (Machingaidze, 1991; Drakakis-Smith, 1984). The war ended in 1980 and Zimbabwe became an independent nation.

The economy that Zimbabwe inherited, though advanced, was underpinned by agriculture, heavily subsidised by South Africa, had a fiscal deficit, high unemployment rate, price controls and a foreign currency deficit (Rakodi, 1995; UN, 2005). At independence, 45 donor countries pledged US$ 1,900 million for development. The Zimbabwe government borrowed large sums on the foreign markets, at least partially in expectation of receipt of aid. When some of the pledges did not materialise, the country was left with a serious debt-servicing problem (Jenkins, 1997).

---

[18] See Phimister 1993 and Machingaidze 1991. The inconsistent implementation of the NHLA meant that, at suspension, some reserves had experienced full, and others partial, implementation of certain provisions of the act. These differences give rise to inconsistencies.

The Zimbabwe government at independence was determined to redress imbalances in economic development in line with expectations of the rural populace that had high expectations of the redistribution of land, had shouldered the burden of the war and formed the political base for the Zimbabwe government (Jenkins, 1997). The government increased investment in communal area agricultural support services and infrastructure and relaxed restrictions on movement (Potts 2000b:879). The government supported the resumption of agricultural activities which had been disrupted by the war. The support extended by the government included a favourable economic environment, increased access to commercial credit, initially heavily subsidised and eventually free government inputs. These included fertiliser, seed, tractor-powered ploughing, extension, irrigation, infrastructure, improved grain marketing, input delivery and veterinary services in communal areas in the first decade of independence. The communal areas experienced an economic boom because of increased support and an excellent 1980-81 rainy season (Berkvens, 1997; Mudimu, 2003). The communal farmers produced 60% of marketed maize in 1982. However, the increase in productivity was limited to 22% of the communal area farmers who produced 80% of all the marketed maize (Bratton, 1987; Palmer, 1990; Breslin, 1994; Amin, 1992). The agricultural output was severely affected by three successive drought years that followed the bumper harvest.

The government introduced free medical services for people earning less than the minimum wage and increased investment in rural health programmes. More than 90% of the population in rural areas was eligible for free health care (Basset *et al.*, 1997:1846). By 1997, 224 rural health service centres had been built and district and provincial facilities upgraded. The increased investment in health services resulted in a three-fold increase in patient attendances at communal area medical institutions. More than a third of pregnant women delivered at formal medical institutions (Loewenson, 1991:367; Marquette, 1997). The increased access to medical facilities and investments by the government in preventative healthcare reduced child mortality, increased life expectancy and increased the general well-being of the communal area populations.

The Zimbabwe government established growth points in an effort to generate employment opportunities in communal areas and halt rural-urban migration. Sixteen centres were accorded growth point status in 1987 (Wekwete 1991:196)[19]. In the study area, Juru Growth Point was established under this programme. Growth points received state support to invest in energy, communications, water supply, and social and administrative infrastructure. Apart from structural investments and

[19] Ideally, growth points were meant to be development sites based on an existing economic base, usually agriculture or mining. The points were envisaged to develop into towns. In the implementation, there was a shift from a focus on economic potential to centres linked to administrative units in the communal lands (Wekwete, 1991:201).

decentralisation of government services, the industrialisation did not take off. At present, growth points function mainly as service depots for passing traffic, with resident populations consisting mainly of civil servants and informal sector workers (Wekwete, 1991:210; Bryceson and Mbara, 2003). The growth points increased opportunities for women to earn non-agricultural income. This included earning cash by serving as waitresses in the bars, shop assistants and prostitutes. Prostitution was found to be an important source of income for younger, unmarried women and older divorcees (Scoones *et al.*, 1996:184; Gregson *et al.*, 1997: 1273).

The number of women working on commercial farms increased as restrictions on movement between European and non-European areas were lifted after the war (Mosley, 1983; Clarke 1977 cited in Adams 1991a:167). In 1999, the Central Statistical Office estimated that 55% of workers on commercial farms were women who were either casual or seasonal employees (Walker, 2006:109; Magaramombe, 2004:51). Women-headed households were more likely to participate in farm wage labour than married women (Adams, 1991a:164). The commercial farms employed women mainly on a temporary and casual basis, which was paid less. As a result, these women had limited access to accommodation and privileges extended to permanent male workers. The lack of accommodation left women vulnerable. They entered into informal liaisons with permanent, mostly male, workers to gain access to accommodation and support between contracts (Adams, 1991a; Rutherford, 2001; Walker, 2006). This increased the number of births to single women, as most of the relationships were seasonal (Walker, 2006:110).

The post-colonial government uses communal areas for urban management as dumping grounds for squatters, vagrants, informal sector workers and prostitutes rounded up in urban areas. Communal areas serve as a scapegoat for the State when it fails to generate employment (Cousins, 1990; Mbiba, 2001). Communal areas have also been used by the government to diffuse the effects of landlessness (Nyambara, 2001b; Cousins 1990). The communal tenure is also used by the State to guarantee 'the right of every citizen to a minimum use of tribal land' (Gluckman, 1945:204). This phenomenon existed in both pre- and post-independence Zimbabwe, illustrating continuities from colonial to post-colonial phases.

The lifting of restrictions on rural-urban movement by the government led to changes in rural-urban connections. Instead of male dominated rural–urban movement, there was an increase in the movement of women and children to urban areas to join their husbands (Potts and Mutambirwa 1990:683). The lifting of prohibitions on African ownership of freehold title in urban areas of Zimbabwe influenced livelihood decision-making on the rural-urban continuum. Africans who had solely relied on communal areas for permanent residence now had an option of owning property in urban areas. While some people shifted the focus of their end of life investment to urban areas and abandoned the rural homestead, others maintained land access and investments in communal areas (Paradza, 2009). The latter was viewed as an important fall-back and insurance against the expensive urban life. The people who invested in urban freehold properties also viewed them as age old security and insurance against communal area risks of drought, failed agricultural production and land shortage. The urban freehold properties provided this security through providing rental income.

The exact nature of rural-urban connections varied between those who continued with cyclical migration of adult males while the women remained to farm in rural areas; some who abandoned the rural homestead and moved a wife and children to the urban areas to join the father and those who for periods of the year had the whole family under one roof in town and the rest of the year (usually the agricultural season) the wife and children returned to the rural areas to engage in agricultural activities. People who were not engaged in wage employment in the urban areas, sent their children to urban areas to take advantage of relatively better education facilities there. Others focused on regular trips to the market to sell their produce, purchase goods for resale or if the distance permitted commuting to work. The differences were informed by gender, distance from the urban areas, financial capital and the way in which prevailing socio-economic conditions impacted on individual domestic units.

3.2.1 Land redistribution

In 1980, the Zimbabwe government embarked on a land redistribution programme[20] which aimed to:

- alleviate population pressure in the communal areas;
- extend the base for productive agriculture in the peasant farming sector;
- improve the standard of living of the largest and poorest sector of the population;
- provide opportunities for people who have no land and employment;
- bring abandoned or under-utilised land to full production;
- expand the infrastructure of economic production and to achieve national stability.

The programme also aimed to redress the inequitable distribution of land between the racial groups. By the end of the programme, less than seven percent of the communal area population had been resettled (Cousins, 1993:31). The then existing legislation restricted the government to acquiring land willingly sold by the European farmers. The programme was abandoned shortly after because the plan was too ambitious and inadequately funded. The Commercial Farmers' Union, a pressure group for the white commercial farmers, successfully lobbied for the slowing down of the resettlement process (Palmer, 1990; Cousins, 1993; Goebel, 2005). The commercial

[20] Zimbabwe, Resettlement Policies and Procedures, Harare, Ministry of Lands, Resettlement and Rural Development 1980 cited in Kinsey, 1999.

farmers offered only marginal, unproductive land for sale to the government, and some of the acquired farms were diverted for personal use by the governing elite (Mudimu, 2003; Goebel, 2005). Furthermore, the post-independence boom in communal area agricultural production misled the government into believing that a sustainable increase in communal area production, capable of meeting welfare needs, might be possible without extensive resettlement (Palmer, 1990:171). The ministries involved were not well co-ordinated and lacked the capacity and manpower to make effective plans for resettlement.

Resettlement had little impact on alleviating land pressure in communal areas. The government of Zimbabwe had planned to resettle 162,000 families when the programme started. At the end of the programme, only 28,600 families had been resettled (Zinyama, 1991:100). The selection criteria for resettlement were biased towards farmers who were formally trained as Master Farmers[21], possessed agricultural implements and were educated and married (Goebel, 2005:89). The financial constraints limited the beneficiary numbers. The conditions for resettled farmers, which prohibited the newly resettled farmers from migrating to work in urban areas, discouraged the people in communal areas from taking up resettlement (Kinsey, 2002). The lack of a clear tenure arrangement between the government and the newly settled farmers also dissuaded some farmers from moving from communal areas (Kinsey, 2002; Goebel, 2005). Some communal area farmers moved to the resettlement areas but retained their land holdings in the communal areas as security (Bratton, 1987:191; Cousins, 1993; Kinsey, 1999). The selection criteria disadvantaged single women, the poor and the landless in the communal areas and had little impact on land pressure in communal areas. Divorced women who left resettlement areas returned to their communal areas of origin (Mvududu and McFadden, 2001; Goebel, 2005). It is not possible to say anything conclusive about the extent to which the land resettlement programme increased communal area hearth-holds' access to productive livelihood resources.

3.2.2 Economic structural adjustment policies (ESAP)

In the 1990s, the Zimbabwe government adopted the Economic Structural Adjustment Policies (ESAP). The policies consisted of a package of individual reform measures worked out in consultation with international agencies and were tied to foreign loans (Norton, 1983; Kanji, 1995). The process involved the adoption of a series of macroeconomic measures including deregulation of the domestic economy, less restrictive trade policies and reductions in public spending (Kanji, 1995; Marquette, 1997). The programme resulted in retrenchments in agriculture, textile, clothing, civil service, leather and construction industries. During the 1991-1992 period,

[21] Master Farmer: a farming qualification which was used to train communal farmers (Potts 2000a:818). The selection criteria favoured men.

21,000 manufacturing and 11,000 health and education jobs were lost (Rakodi, 1994a, 1994b cited in Potts, 2000b:885). ESAP increased unemployment rates through retrenchment and failure by school-leavers to secure employment. The total number of school-leavers seeking jobs in 1990 was in excess of a million, whereas the job creation target in the Zimbabwe government's First Five Year Plan was 144,000 (Drinkwater, 1991 cited in Nyambara, 2001b:267). These figures do not include casual and seasonal labour which were sectors dominated by women (Kanji, 1995:38). The urban poor suffered as the loss of wages was coupled with increases in costs of urban services that followed the removal of subsidies (Potts and Mutambirwa, 1998; Rakodi, 1995). The informal economy workforce in the urban areas of Zimbabwe rose from 10% in 1980 to an estimated 40% in 2004 (UN, 2005). This resulted in a decline of remittances to communal areas. The way in which the destabilisation effects of structural adjustment programmes simultaneously affected both rural and urban areas highlighted the complementary relation between the urban and rural sectors in Zimbabwe.

Communal areas were vulnerable to ESAP because of their dependence on wage employment for cash flows and agricultural inputs. This dependence was created during the colonial era and perpetuated by the Mugabe government's post-independence policies. Communal area farmers also suffered from food price increases, as they were net food purchasers (Potts, 1995; Rakodi, 1995; Alderson, 1998). Prices of agricultural inputs, social services, agricultural equipment and livestock dipping[22] increased significantly (Bird and Shepherd, 2003; Potts, 2006). The free provision of agricultural inputs was converted to food-for-work and loans (Bratton, 1987). The decline in investment in rural infrastructure such as roads also negatively affected rural cash crop growers' access to the markets (Alderson, 1998). ESAP disrupted communal areas' food production as input supplies became erratic. Cash shortages constrained purchasing power, and earnings from crop sales declined. The removal of subsidies and price control on public transport following the structural adjustment programmes resulted in huge transport cost increases which had a negative impact on mobility. Since the capacity to engage in agricultural production depended on access to cash generated from wage employment, ESAP weakened the rural cash flows, with a consequent negative impact on agricultural production. This further undermined the capacity of single women to produce food for their own consumption and for sale. The development also reduced the incidence of casual work as a coping strategy, as the main providers of such work also suffered from the income shock. Economic Structural Adjustment Programmes may have created new opportunities, as marginal niches (such as lower wages at commercial farms) became more numerous, and typically women occupy these niches.

[22]Livestock, usually cattle, are dipped frequently in chemicals to prevent sickness. Previously, this service was provided by government. After ESAP, individual farmers had to purchase their own dipping chemicals directly from the suppliers with no subsidy from government.

The government reduced health spending by a third and re-introduced user fees. The reduced public expenditure led to a decline in outreach activities, drug shortages, declining clinic attendance, increased infant mortality and loss of staff due to the brain drain (Basset *et al.*, 1997)[23]. This further undermined communal area livelihoods. The decline in health services increased women's burden of care, which traditionally is gendered. The incapacitated and/or ill people returned to the communal areas where it was cheaper for them to be cared for under the traditional support systems which relied on women to take care of the ill.

After ESAP, the government monopolised control of agricultural produce and reintroduced price controls in an effort to maintain low food prices for a discontented urban population. The reintroduction of price controls discouraged both commercial and communal farmers from growing maize, resulting in a net shortage of maize (Bird and Shepherd, 2003). The combined effects of ESAP and government price controls resulted in erratic and inappropriate supply of agricultural inputs in Zimbabwe (Ncube, 2004). This has undermined commercial and subsistence agriculture. As a result, communal area farmers have become increasingly vulnerable as they simultaneously lost the capacity to produce subsistence food, the option to buy from the market, and were no longer receiving remittances from the urban areas.

ESAP divided some families again as wage labourers' careers shortened and urban incomes dwindled. Retrenched workers returned to take up their residence in the communal area homesteads which they had maintained as security against such eventualities. Another response was the increased movement of people from rural areas to seek work in the urban informal sector to replace the agricultural income that had been curtailed by a reduction of remittances. The structural adjustment programmes induced change in communal areas by significantly altering the dynamics of mobility patterns. The movement was not systematic. In some instances, contraction of urban employment opportunities resulted in return migration, which caused tensions in rural areas as there were inadequate agrarian and non-agrarian opportunities to absorb the returnees. Some people reacted to the impacts of ESAP by migrating to urban areas, as they were no longer able to farm. This mainly involved the movement of women and children joining their husbands as a migrant strategy to reduce costs by having a single residence (Bryceson and Mbara, 2003; Ranga, 2004). Most of the migrants retained their rural land links as security against increased urban insecurity.

The ruling party, ZANU-PF, lost its popularity in the 1990s due to the economic downturn, currency collapse, food shortages and unemployment (McGregor, 2001; Hartnack, 2005:178). In 2000, a new political party, the Movement for Democratic

[23] The health staff that left Zimbabwe went to Botswana, Australia, United Kingdom, New Zealand and South Africa.

Change (MDC), was formed. ZANU-PF began a process of violent campaigns and intimidation aimed at undermining the MDC and bolstering ZANU-PF's rural base (Alexander and McGregor, 2001; Hartnack, 2005). The process resulted in the intimidation and politicisation of access to productive resources in the rural areas. This process introduced new arenas for negotiating resource access in communal areas. These weakened the existing mechanisms of productive resource access and introduced vulnerabilities and opportunities for different groups of people.

3.2.3 'Fast track' land resettlement programme

In order to deal with the growing tensions resulting from these austerity measures and economic decline, the government launched the fast track land resettlement programme in 2000. The programme was initiated to woo support for the ruling party, which had dwindled (Hartnack, 2005:177). The programme resulted in the forced transfer of land from European commercial farmers to African landowners. This had a negative impact on the productivity of these farms. The people who benefited from the fast track land redistribution effort maintained their rights to communal land. The Zimbabwe government was not able to provide financial and infrastructural support to the people who resettled on the former commercial farms (Mudimu, 2003; Magaramombe, 2004; Goebel, 2005; Hartnack, 2005). Some of the people who left the former commercial farms have returned to the communal areas because of lack of government support for the new farmers, lack of infrastructure such as schools and health facilities, and forced evictions by government officials (Sithole *et al.*, 2003).

The fast track land resettlement resulted in the loss of permanent and casual commercial farm employment opportunities, which had formed an important source of communal farmers' off-farm income (Mudimu, 2003; Hartnack, 2005). An estimated 80,000 workers, who with their families made up half a million people, were affected (Sachikonye, 2003). The fast track land reform programme resulted in a loss of livelihood and in many cases a place of residence for former commercial farm workers (Mudimu, 2003; Magaramombe, 2004; Hartnack, 2005; Walker, 2006). The Zimbabweans returned to their communal areas, with 46% of displaced workers settled in communal areas of Mashonaland East, and some remained on the commercial farms (Magaramombe, 2004:48). Stranded women resorted to commercial sex with influential people to gain access to accommodation and land in communal areas, growth points or commercial farms (Walker, 2006:111).

The disruption of commercial agriculture had a negative impact on the country's industry, which was largely agro-based. The closure of seed producing farms as a result of the fast track land reform disrupted the country's maize seed production. The shortfall had to be imported (Ncube, 2004).

The fast track land distribution resulted in targeted sanctions from the West. The negative publicity affected the tourism industry, and the economy continued to deteriorate. In January 2005, the Zimbabwean economy was characterised by a high inflation rate, shortage of foreign currency, inadequate investment, a budget deficit and stagnating employment. The shortage of foreign currency restricted access to supplies of essential imports needed for industrial and agricultural production, fuel and basic commodities. The erratic supply of fuel severely affected agricultural production, mechanised draught on commercial farms, and the distribution and collection of inputs in rural areas. The declining formal economy had a negative impact on the government's revenue base (UN, 2005). The disruption of commercial farming destabilised Zimbabwe's already unstable revenue base through loss of taxes, foreign currency and food. Historically, commercial farming areas had been responsible for growing the food that communal and urban area residents purchased on the market in drought years. After the fast track land reform, there was a shortage of grain on the market. This forced communal farmers to increase reliance on their marginal soils and state transfers of food. The government's ability to import food was restricted by its limited foreign currency. The commercial farm workers who moved to the communal areas increased demand for land, and this increased the diversity of ways that people gain membership and access to residential land in these areas. The disruption of commercial farming increased the vulnerability of single women by reducing the options available for them to diversify into commercial farm work. This forced women to diversify into riskier activities.

3.2.4 'Operation Restore Order'

In May 2005, the Zimbabwe government implemented operation Restore Order. The motives behind the operation included a general concern felt by the government to do something about the chaos and congestion that characterised urban areas and to check the parallel foreign currency market. The operation was also believed to be a retribution for urban dwellers as they had voted for the opposition party in previous elections. Its aim was to disperse selected urban populations to rural areas where the ZANU-PF party hierarchy could more easily control them. An alternative view saw the operation as a pre-emptive strategy to check popular uprisings in light of the deepening food insecurity, as a diversion of attention away from the under-utilisation of land on the commercial farming areas and as a way to stifle independent economic and political activity in the country's urban areas (UN, 2005; Bratton and Masunungure, 2006). The operation led to the destruction of people's homes, sources of livelihood, shelter, access to services and business premises, and property in urban areas and growth points. The UN estimated that some 700,000 people were directly affected. Of these, 40,800 families were headed by women. The operation also indirectly affected other people through the loss of rental income, disruption of networks involved in the supply chain of the informal sector, credit institutions and casual and part-time labour. Some estimated 114,000 people returned to the rural areas (Action Aid International, 2005; UNAIDS, 2005; Bratton and Masunungure 2006; Potts, 2006). The operation disrupted livelihood productive resource access chains for some communal area based people. The displacement of people from urban to rural areas reactivated people's interest in communal area homesteads which once again provided a refuge from 'hardships of urban life'. The disruptive impact of operation Restore Order on rural-urban connections cannot be generalised here but will be explored in more detail in subsequent chapters.

3.2.5 HIV/AIDS and mortality in Zimbabwe

In this context of economic decline and political tensions, the HIV/AIDS epidemic has added another burden on the population of communal areas. An estimated 34% of the Zimbabwean population was infected with HIV at the end of 2002 (UNDP, 2004a:11). The 20-49 year age group accounted for more than 70% of AIDS cases. This is the most economically active group, and their morbidity and mortality was a significant loss of human capital. The deaths increased dependency ratios as mostly the elderly and children were spared from the epidemic (NACP, 1998; UNAIDS, 2003; Ministry of Health and Child Welfare, 2004). In a reversal of earlier trends, more women than men were now being infected, with 28% among women compared to 19% among men (Gregson *et al.*, 2002). Widowed and divorced women were more significantly associated with higher HIV infection than married and never married women (IDS, 2003).

A 2005 spatial analysis of the HIV/AIDS epidemic in Zimbabwe showed that urban area prevalence was 28%, rural areas 21%. The 'other' category that included commercial farms, growth points, army camps and mining areas had a prevalence of 35%. 'Other' areas were set apart because they have a much higher HIV prevalence than traditional rural areas (IDS, 2003). A study in rural Zimbabwe concluded that HIV prevalence was higher at roadside trading centres along major highways (Gregson *et al.*, 2002). The commercial farms, growth points, army camps and mining areas contained no more than 10% of Zimbabwe's population (Ministry of Health and Child Welfare, 2004; UNDP, 2004b). The high vulnerability of commercial farming worker populations is explained by the lack of decent accommodation for workers on the farm, lack of recreational facilities and lack of social control induced by the multi-ethnicity of the populations (IDS, 2003). The close proximity of Goromonzi communal area to the 'other' areas rendered its population more vulnerable to the pandemic than other communal areas.

HIV/AIDS has increased the number of single women by increasing the numbers of widows. Premature male adult mortality has left young widows, many of whom have not had the necessary time to build up a set of extra-family levers such as access to community land, community groups and micro-credit which can be used to exert power within a family (Haddad and Gillespie, 2001; Drinkwater, 2003). HIV/AIDS has destabilised already stressed social support networks and put further pressure on unstable dependency relationships across generations and between men and women (Walker, 2002; Drinkwater, 2003). The stigma associated with the HIV/AIDS epidemic has been found to exclude widows from accessing some informal networks and survival strategies, such as working for other people (Walker, 2002). The increased awareness of HIV/AIDS has led to a gradual decline of the widow inheritance system and has deprived widows of an important survival strategy (WLSA, 1994).

Higher death rates have led to increased accusations and pre-occupations with witchcraft (Drinkwater, 2003). Widows and women who have lost children are more vulnerable to witchcraft accusations. Widows are more likely to be blamed for the death of their husband and therefore asked to return to their natal family, resulting in the loss of land access (Schmidt, 1992; Vijfhuizen, 2002). Research carried out in Southern Africa and Kenya concurred that HIV/AIDS was worsening the already vulnerable position of women and children. Land-related disputes had increased, as had tension over land allocation to, and the rights of, widows. There was, however, no evidence of distress land sales. The studies noted reduced land-use and changes in cropping patterns by affected households due to labour shortages and lack of income for agricultural inputs (Yamano and Jayne, 2004; Aliber *et al.*, 2004; Drinkwater, 2003).

As in the colonial times, the communal area subsidised the urban area by producing and reproducing labour. HIV/AIDS and the declining state health system have had an enormous impact on rural–urban connections. Zimbabwean rural areas bear the larger burden of the morbidity caused by the HIV/AIDS epidemic because ill people generally return to the communal areas where life is cheaper and where the immediate and extended family is present to provide care (Mutangadura, 2000; Andersson, 2002). The reduced capacity of the Zimbabwe government to provide healthcare has led to the promotion of the practise of home-based care for HIV/AIDS patients. Home-based care services are intended to provide support to families and patients in the home and reduce the need for hospital admission. Although the initiative has many benefits, the levels of coverage are low and many patients at some phase of their illness need more than their home can provide (Jackson and Kerkhoven, 1995; Hansen *et al.*, 1998). Home-based care has increased the burden of women who are the majority of the caregivers of the afflicted. Carers spend up to three and a half hours a day on routine patient care (Hansen *et al.*, 1998:751). The increasing prevalence of HIV/AIDS among women has resulted in an increase in orphaned children, most of who are cared for by elderly women in grandmother-headed households. Ill single women are specifically more vulnerable because the illness reduces the woman's capacity to provide for her hearth-hold.

3.3 The communal areas as a specific type of governance space

The governance system in communal areas has influenced the opportunities for women to access productive resources. It is not possible to give general trends here. The repeated government interventions and local conditions in specific communal area have resulted in specific types of productive resource governance at local level. The colonial and post-colonial governments maintained land tenure as communal in order to retain political control over productive resources especially land and to prevent the residents from becoming economically independent (Cousins, 1990; Cheater and Gaidzanwa, 1996; O'Flaherty, 1998). In this way, the State controls the rural population and is able to confine it to a specific space. An example is the way the Zimbabwe government uses food and drought relief to mobilise votes during elections (Mudimu, 2003). The term 'customary' is also used to refer to communal areas as a legal sphere in which traditional and/or customary laws are observed under the administration of traditional authorities who include chiefs, sub-chiefs, village heads, traditional authorities and spirit mediums. The chieftainship is divided into smaller units headed by sub-chiefs or headmen. These are divided into smaller villages headed by a *sabhuku*/kraalhead/village head (Holleman, 1969; O'Flaherty, 1998; Andersson 1999). This thesis uses the term *sabhuku*. The study villages of Makuku and Ndamba both fall under Chief Chikwaka's domain in Goromonzi.

The colonial government reorganised chieftainships (Ranger, 1982; Lan, 1985). During the NLHA, the State, through European Native Commissioners, assumed responsibility for the allocation and control of land in the communal areas but did not wholly override the practise of land allocation and control by chiefs and lineage heads (Holleman, 1954; Weinrich, 1975; Ranger, 1982; Cheater, 1986; Machingaidze, 1991). The 1967 Tribal Trust Act recognised the traditional leaders and the traditional tenure systems' control over land and gave them the authority to allocate land (Cheater, 1990; Ranger, 1993; Scoones *et al.*, 1996). The State used these traditional leaders to administer a codified customary law, enforce land conservation policies and offer propaganda advantage to the colonial government (Ranger, 1982; Bell and Hotchkiss, 1991). During the war, the traditional leaders lost their legitimacy with the Africans and guerrillas as they were viewed as agents of the colonial government (Ranger, 1982; Andersson, 1999; Vijfhuizen, 2002). Ultimately, the traditional authorities' authority was based on their accountability not towards their citizens, but towards the State. The establishment of protected villages[24] and the control by guerrillas in other communal areas further undermined the influence of traditional authorities during the war (Ranger, 1982; Ladley and Lan, 1985; Kesby, 1996; Andersson, 1999).

[24] Guarded and fenced villages established during the war by the Rhodesian government in an attempt to assert control over rural society (Kesby, 1996:54).

At the end of the war, there was a vacuum of formal authority in the communal areas, and people claimed land for cultivation previously reserved for grazing (Ranger, 1993; Kesby, 1996; Scoones *et al.*, 1996). The agricultural boom of the 1980s generated intense competition for land in some communal areas. In Gokwe, land was acquired through political connections, individual initiatives and outright purchase as farmers competed to increase output. The powerful gained the most through land grabbing and conversion of community land held in trust for individual use (Nyambara, 2001a).

The post-independence government created parallel structures to customary authorities in addition to the administration that existed before independence. This led to a proliferation of governance institutions in communal areas. The independent government of Zimbabwe initially marginalised the traditional leaders because of their alleged collaboration with the colonial government. The 1982 Communal Lands Act gave Rural District Councils the power to allocate land in communal areas (Brand, 1991; O'Flaherty, 1998; Nyambara, 2001a). The 1984 Prime Minister's decentralisation directive was adopted, which set up the Village Development Committee, Ward Development Committees and the office of an elected councillor in place of traditional leaders at the local level (Brand, 1991). The ZANU-PF party structures from the war also asserted their authority and took over a number of functions that had formerly been exercised by chiefs, headmen or the administration (Brand, 1991:90). The chiefs continued to allocate land, claiming to defend tradition against state policies, and this increased the conflict between traditional and elected leaders (Vijfhuizen, 2002; Andersson, 1999).

In 1992 the Government of Zimbabwe formally restored the traditional leaders' authority over land allocation and management (Cheater, 1990; Cousins, 1993; Ranger, 1993; Scoones *et al.*, 1996; Nyambara, 2001a). The fast track land reform process put in place new political institutions and actors, including war veterans and ZANU-PF youths, who were able to exert a certain degree of power over representatives of the State and traditional authorities (Hartnack, 2005; McGregor, 2001. Although the fast track occurred in the commercial farming areas, the process reinvented the war veterans and ZANU-PF party youths in the communal farming areas. The political actors and institutions assumed administrative roles in addition to the traditional and state authorities. The result is that chiefs, headmen, spirit mediums, traditional healers, ZANU-PF party leaders, entrepreneurs, village committees, peasants and households are now all involved in struggles over land allocation and management in communal areas (Lan, 1985; Cheater, 1986; Cousins, 1993; Andersson, 1999; Nyambara, 2001a). With the exception of spirit mediums, these governance structures have a male bias. The introduction of multiple authorities into communal areas by successive governments has produced a complex governance system. This means that neither patriarchy nor any other system is dominant. This has created uncertainty and paradoxically also increased land access opportunities for a variety of groups such as single women and former commercial farm workers who were marginalised by the traditional communal area land governance systems. However, after all the transformations, which the communal areas have undergone, it would be inaccurate to continue to label the governance and land tenure system as customary.

The nature of the community in the communal areas is widely contested: there are those who see the communities as cohesive (Holleman, 1969) and those who challenge the notion of cohesive community. The latter cite the existence of widely contested and at times overlapping village boundaries as proof (Ranger, 1982; Cousins, 1990, 1993; Andersson, 1999). Villages serve as a formally recognised administrative unit in the communal areas. The evolution of the villages and their composition provides an insight into the nature of the communities in the communal area villages.

Villages were created during the colonial period and organised under the office of *sabhuku*/kraal head who was responsible for collecting the taxes for the colonial government (Holleman, 1954; Weinrich, 1975; Ranger, 1982). As already stated, during the war, some villages were reorganised as protected villages set up by the Rhodesian government while the guerrillas controlled the others (Ranger, 1982; Kesby, 1996). The ZANU-PF party consolidated the colonial villages under the ZANU-PF village committees at independence. The Zimbabwe government uses the villages as administrative units for distributing food relief and collecting taxes. The governance of productive resources within the villages remains unclear. There are frequent inter-village land transfers, grazing and forestry resource sharing and social security arrangements. This has increased the complexity of governance structures.

The foregoing illustrates the multiple, plural, legal governance institutions that govern land, labour and forest use in communal areas. As a result, the authority and the legal guidelines of traditional leaders have become unclear, confused and in many cases overlap with those of other institutions governing natural resources, land, labour, village membership and government agricultural inputs and labour opportunities access. The contradictory findings in the literature arise because of the differential interpretation and application of rules by the various administrators and the specific conditions in each communal area. As a result, the various opportunities, constraints and outcomes of these governance systems on specific locales will vary among the 176 communal areas and the villages within them.[25]

It is within these systems that there is supposed to exist a system of property rights in which individual households within a group or community enjoy rights of use over specific land parcels for cropping and residential use, and a defined group or

[25] The present study is unique in that it discusses a unique set of conditions in Mwanza communal area. However, at the level of processes and multiplicity of land tenure systems, it is a reflection of the general situation in Zimbabwe's communal farming areas.

community has rights of access to an area of common land which provides grazing, fuel, wood, water and building material (Cousins, 1993). Individuals are supposed to gain land through membership of the community. The system allocates primary land rights to men and secondary land rights to women. Under this system, daughters and widows cannot hold land independent of a man. The literature is contradictory on the security of the customary right to land (Pankhurst, 1991; Berkvens, 1997; Andersson, 1999). The contradictory positions are a reflection of the different ways in which the various governance institutions interact with specific local conditions in communal areas of Zimbabwe.

However, under pressure of all kinds of changes in the context of communal farming areas, new ways of allocating land have emerged in practice. The traditional systems were transformed by interventions of native commissioners allocating land during the NLHA, guerrilla and colonial government forces during the war, and the re-organisation of local government after independence. The changing rural–urban connections in response to the war, land redistribution and wage contraction has affected the character of the communal areas. The movement of former commercial farm workers into communal areas has increased the demand for land, spawned land markets and increased the diversity of the populations. The nature of rural-urban connections varies with the location and governance systems of a communal area. The developments introduced negotiating space within which single women hearth-holds could gain access to land, community and natural resources in the communal areas. All these deviations from traditional norms have increased the space within which single women can negotiate access to land in communal areas. The specific local conditions are important for understanding the exact impact of the multiple governance institutions on people's daily lives. Single women are also a heterogeneous group who experience conditions differently. The focus on individual hearth-holds and locations will help to increase the understanding of their experiences.

3.4 The evolution of livelihoods in communal areas

In the early colonial era, when agriculture was the mainstay of existence, elderly males controlled the means of production, land, cattle and marriage. Young men relied on their patri-kin for land and bride price, and nearly all marriages were arranged by parents. Women's consent to marriage was not required (May, 1983:21; Folbre, 1988). The colonial government increased the economic dependence of African women on men by giving males the control of all forms of property, including women in communal areas. This was achieved through the registration of rights to communal land in men's names under the Native Land Husbandry Act under the pretext of customary law. Women were confined to communal areas during the colonial period (Barnes, 1997; Moyo and Kawewe, 2002; Shutt, 2002). The movement of African women from communal areas was restricted by denying them access to urban housing and restricting the use of motorised transport to African women who had written permission from their male guardians (Schmidt, 1992; Cheater and Gaidzanwa, 1996; Barnes, 1997).

The migrant labour system favoured the employment of single male migrant workers, through 'pass' laws and the provision of urban residence only for single males. The Europeans educated African men and not women, because the men were expected to come into contact with Europeans through labour migration. The communal areas were maintained for the subsistence of workers' families and the support of men past their working age. Consequently, labour costs were subsidised by rural subsistence production, which bore all the cost of the production and reproduction of labour and maintenance of workers when they were ill, disabled or old (Potts and Mutambirwa, 1990:678). The development left women with no livelihood option but to farm. The poor soils and ecological environment made sole dependence on agriculture in communal areas risky. Single women were more vulnerable as they were more dependent on agriculture and often had no male relatives to support them with cash from wage labour.

Urban wage employment destabilised traditional patriarchal relations by freeing young men from agricultural livelihoods and gave them an opportunity to make more decisions about marriage without their father's consent. This also made dissolution of marriage through divorce easier (May, 1983, 1987). The destabilisation of patriarchal relations undermined the seniority and control of the elderly and increased their insecurity and dependency on the young. Over time, peasant farming became a fairly low status activity, left to women or retirees. The women who remained in the rural areas became the *de facto* heads of households (Cheater and Gaidzanwa, 1996; Barnes, 1997; Potts, 2000a). The separation of married people changed gender dynamics, as high rates of male migration led to increased adultery, new forms of informal unions and prostitution (Goebel, 2005:36). All these processes contributed to the increase of the number of single women in communal areas. The out-migration of males and women's limited access to livelihood opportunities outside the communal areas resulted in the feminisation of communal areas (May, 1983; Schmidt, 1992). The result was that male-dominated agriculture ceased to be a dominant form of livelihood in the communal areas, although the presence of elderly retired men and the continued return of men to the communal areas limited the feminisation of agriculture and communal area residence. The continued dominance of males in the land, village membership and natural resource governance systems perpetuated the patriarchal control over land and other productive resources. The weakening of patriarchy may have opened space for women's autonomy; however, this development has also increased the vulnerability of single women as it has undermined the institutions that obliged men to take care of them.

The post-independence government of Zimbabwe made a series of legislative changes aimed at changing gender relations. A Ministry of Community Development and

Women's Affairs was established in 1981. The Legal Age of Majority Act was passed, giving all races and genders full contractual rights on attaining the age of 18. The Law and Primary Courts Act allowed primary courts to take decisions that had previously been determined by chiefs under traditional law. The passage of the Matrimonial Causes Act granted women the right to part of the marital property in the event of a divorce (Batezat and Mwalo cited in Hindin, 2002:154). In 1997, the Administration of the Estates Amendment Act ended customary inheritance and allowed widows to inherit the house, land and property of their husband (Kesby, 1999:38). Although the legal reforms freed women from legal disadvantages of perpetual minority and reduced the authority of patriarchs over women (Rwezaura, 1989: 15; Hindin, 2002), the legislation had no impact on women's access to land in the communal areas. The post-independence government maintained traditional communal tenure and supported the administering of customary law on inheritance in the communal areas (Kesby, 1999:38; WLSA, 1990) because of the ZANU-PF party's desire to maintain the political support of men (Kesby, 1999). As a result, men have maintained ownership of land and homesteads in the communal areas.

The vulnerability of communal area livelihoods, which started when they were established, has been heightened over the years by drought, withdrawal of able-bodied male labour, disruption of commercial farming, economic stagnation in the modern sector, structural adjustment programmes and HIV/AIDS which have all induced a constant shift in rural-urban connections. The reduction in opportunities for diversification caused by structural adjustment policies, economic crisis, land reform and operation Restore Order increased the strain on the already marginal livelihoods of communal area inhabitants. Drought is a recurring phenomenon in Sub-Saharan African semi-arid zones (Scoones *et al.*, 1996:4). The frequent droughts threaten farmers' ability to meet their own food requirements and induce changes in people's livelihoods (Cousins, 1993; Scoones *et al.*, 1996; Kinsey, 2002; Bird and Shepherd, 2003). The coping strategies used over successive droughts demonstrate a decline in the ability of communal areas to produce subsistence crops, their increased reliance on external links and changes in the nature of engagement with the urban sector. A survey of literature on drought experiences shows a gradual shift from localised responses to increasing dependence on the areas outside communal areas, including the State, the market and wages (Scoones *et al.*, 1996:4; Kinsey, 2002; Bird and Shepherd, 2003).

The pre-colonial people coped with drought by collecting wild food and grain exchanges. The coping strategies changed during the colonisation period to include the exchange of livestock for grain and provision of drought relief by the colonial government (Drinkwater, 1989; Scoones *et al.*, 1996). In the 1920-1940 period, the use of cattle sales as a strategy declined because of the reduced number of cattle and smaller herds brought on by increased human population and the de-stocking programme promoted by the Native Land Husbandry Act. Wages became an increasingly important source of normal and drought-time support for communal area livelihoods (Ranger, 1982; Scoones *et al.*, 1996). In the post-independence period, communal areas increasingly struggled to meet consumption demand even in non-drought years. Farmers had to resort to purchasing from the market or food remittances from urban areas, to meet the food deficit that occurs just before harvesting. There was increased dependence on the government and migrated relatives during the 1982-84 droughts. The post-independence government initially offered drought relief, which was replaced with food-for-work programmes (Amin, 1992; Scoones *et al.*, 1996; Kinsey *et al.*, 1998). The drought decimated communal cattle herds, impairing the capacity of households to recover in the long term (Alderson, 1998). The reduction in the livestock undermined the communal area inhabitants' attempts to meet their tillage needs. This reduced agricultural production.

The decline of agricultural production led to an increase in non-agricultural activities in the communal areas (Bryceson, 1999, 2002; Andersson, 1999). There was reduced usage of arable lands in the affected communal areas. The land under fallow increased tenfold in communal farming areas in 2000 (CSO 2000 Agriculture Livestock Survey in communal areas cited in Mudimu, 2003:19) because of soil exhaustion, the high price of fertiliser which made longer fallow the only affordable soil management practice, the shortage of labour and draught, illness or disease (Bird and Shepherd, 2003; Mudimu, 2003: 19). The under-utilisation of agricultural land in communal areas confirmed the declining use of land as an agricultural asset and the increasing importance of livelihood diversification. The changing land use and demand patterns cannot be generalised to all the communal areas of Zimbabwe, nor is it correct to assume that the 1980s were the same as the 1990s. This finding contradicts descriptions of communal areas as overcrowded and land short (Cheater, 1986; Machingaidze, 1991; Phimister, 1993; Scoones *et al.*, 1996; Andersson, 1999; Thompson, 2004). The two opposing views highlight the extent of disagreement by scholars on the land access situation in communal areas. Each of these situations implies different livelihood opportunities for single women.

Communal area residents' employment options arise from agricultural, non-agricultural activities and through mobility. Agriculture-related employment is derived from work on other people's land in communal areas and co-operative work. The non-agricultural activities that people in communal areas increasingly rely on for livelihood generation include casual labour, trading, beer brewing, craftwork and migration. Access to these opportunities is gendered. Casual labour refers to non-permanent, daily paid work performed for a variety of employers (Adams, 1991b:298). The work is remunerated through a range of payment arrangements including cash, payment in kind or a service like ploughing. Casual work is performed by both men and women. Men dominate the ploughing and the non-agricultural jobs (Adams, 1991b:311; Jackson and Collier, 1991:39; Scoones *et al.*, 1996:183).

The households that hire casual labour are wealthy and have high crop sales (Adams, 1991b:313; Pankhurst, 1991:614; Berkvens, 1997:16). Female-headed households are more likely than male-headed households to hire wage labourers (Adams, 1991a:164). Households headed by a casual worker do not in general hire workers, sell any crops or receive regular sizeable remittances. The households that hire out labour are mostly poor and headed by single women (Richards, 1939 cited in Adams, 1991a:165, 1991b:316, 318). This shows that single women headed households are represented among the wealthiest and poorest domestic units.

Permanent workers perform general agricultural labour or serve as shop assistants, barmaids or domestic house-help. The permanent workers normally reside at their place of work (Adams, 1991b; Jackson and Collier, 1991). More than two thirds of the permanent workers are male. Female permanent workers raise children, cook, clean and fetch water. They are generally aged between 16 and 30 years. Shop owners in the communal areas prefer to hire divorced or widowed women to work in their stores or beer halls because the women are willing to work for less money and are more dependable (Adams, 1991a:175). Another source of waged work is live-in domestic help in urban areas. A study of urban domestic workers in Botswana concluded that the expensive urban environment and conditions of service made it difficult for these women to reside with their children (Izzard, 1985).

Some communal areas are close to commercial farms where residents used to have access to temporary or permanent work in agriculture. The number of women employed on commercial farms has increased steadily since the colonial era. The figure continued to rise after independence as women had to support retrenched men (Adams, 1991b; Barnes, 1997; Jackson and Collier, 1991). The availability of commercial farm employment was curtailed by the fast track land reform (Chadya and Mayavo, 2002; Magaramombe, 2004; Goebel, 2005; Hartnack, 2005). The fast track land reform prompted a return of commercial farm workers to the communal areas. Their return increased competition for land, opportunities for non-agricultural employment and transfers from the State.

Women and households with lower remittance receipts dominate trading in communal areas. They trade in vegetables, beer, handicraft, pottery and bricks (Pankhurst, 1991; Cousins, 1993; Scoones *et al.*, 1996; Berkvens, 1997). Trading is hampered by the low availability of cash among the communal area populace, stiff competition and erratic availability of goods. The government has restricted trading through strict licensing and regulation of all commercial activities in rural areas. For example, the preparation and sale of food was, and is, still restricted to licensed premises[26] at growth points and established commercial centres (Cheater and Gaidzanwa,

[26] Illegal traders were harassed by the police, had their goods confiscated and were ordered to pay a fine. This undermined the viability of the activity (Muzvidziwa, 2000:82).

1996; Scoones *et al.*, 1996; Muzvidziwa, 2000). A study on a growth point showed that, though single women were the majority of the traders, they had restricted access to licensed trading space. The available space was allocated through existing governance institutions which favoured men. Single women were discriminated against in the allocation of trading space because of their marital status (Zhou, 1993 cited in Gaidzanwa, 1997: 162-163). The option of cross-border trading is a highly lucrative activity closed to the majority of women from rural areas as they lack cash and the formal documents required for these activities (Cheater and Gaidzanwa, 1996; Muzvidziwa, 2000, 2001).

Operation Restore Order severely disrupted the informal economy by disrupting networks, destroying businesses and displacing the informal sector workers. This reduced women's alternatives for livelihood diversification and increased their reliance on risky activities. Unmarried women and older divorcees used prostitution and temporary liaisons with men to gain access to resources (Pankhurst, 1991:623; Scoones *et al.*, 1996:184; Gregson *et al.*, 1997:1273; Mudimu, 2003:35; Hartnack 2005:183). Prostitution, one of the diversification options for women with lower asset levels, is concentrated around growth points, rural business centres, beer parties and commercial farming areas. The limited options for single women to diversify their likelihoods and the risks associated with agriculture have increased their desperation. The single women who engage in transactional sex are vulnerable to HIV/AIDS and giving birth to children they cannot support.

Commercial and ritual beer brewing activities generate opportunities for non-agricultural employment. Women dominate beer brewing, an activity concentrated around elderly people's homes (Helmsing, 1991; McCall, 1996; Scoones *et al.*, 1996; Vijfhuizen, 2002). Historically, beer has been used as payment for working groups and for specific time-bound labour activities such as house construction, land clearing or agricultural work parties. In order to conduct successfully a beer brewing business, women have to make a substantial capital injection to purchase grain, firewood and labour for transporting water. Beer brewing provides more income than other businesses (McCall, 1996). Traditional beer is also used appease spirits (Vijfhuizen, 2002:2210).

Pottery is the specialist activity of a few women (Helmsing, 1991; Scoones *et al.*, 1996:185). Skilled women make domestic utensils for sale to the local population. Pottery is also vulnerable to the volatility of rural markets and seasonality. Potters cannot manufacture pots during the rainy season as they rely on the sun for part of the drying process.

Originally a male dominated activity, migration has increasingly been adopted by women as they take up employment in urban formal and informal sectors (Drakakis-Smith, 1984; Barnes, 1997; Bryceson and Mbara, 2003; Ranga, 2004). The easing of laws restricting movement from communal to urban areas by the post-colonial government has increased female mobility. The need to balance cash earning with domestic roles restricts women's mobility options and increases their vulnerability to the weak and volatile communal area market. Even then, the historical analysis of the nature of women's rural-urban connections shows that the connections have evolved significantly. Initially women were confined to rural areas and the extent of their connection was the rearing of children, farming to subsidise the urban sector and as the recipients of remittances from the waged male migrants. After restrictions on mobility and urban land ownership were lifted, women's connections evolved as they increasingly traversed the rural-urban continuum working, trading and contributing to the development and maintenance of both rural and urban bases. It is not possible to generalise women's participation in rural urban connections. Single women mediated rural-urban connections differently compared to women who were in a conjugal relationship.

Remittances are an important source of livelihood diversification. Remittances are also important for investment in agricultural activities and a buffer during drought (Jackson and Collier, 1991: 37; Scoones *et al.*, 1996). The main recipients of remittances are married women and the elderly. Although single women receive fewer remittances than other household heads, the remittances enable the receiving households to invest in agriculture, casual and permanent labour that are important livelihood activities for single women. Remittances have declined since the increased cost of living in urban areas following the structural adjustment, retrenchments and the contraction of the economy (Scoones *et al.*, 1996; Mudimu, 2003; Huisman, 2005). HIV/AIDS has reduced remittances, as the most vulnerable are the economically active members of the population. The value of money has also been eroded by the high inflation that characterises the Zimbabwean economy.

Although agriculture is no longer the dominant livelihood activity, poorer households continue to farm because these households lack resources to diversify their livelihood portfolios (Berkvens, 1997; Andersson, 1999). Destitute households increase their dependence on networks of social relations, the State and the church (Scoones *et al.*, 1996:187; Mudimu, 2003). The state drought relief in years of severe drought was erratic but remained vital for many desperate households (Scoones *et al.*, 1996; Mudimu, 2003). Households that fail to cope migrate or succumb. The pathways of different households demonstrate that experiences depend on the household's access to productive resources.

3.5 Implications for single women's livelihood decisions and productive resource access in the communal farming areas

Single women in communal areas of Zimbabwe make livelihood decisions and access productive resources under very difficult circumstances. This is because of the historic and recent developments in the communal areas. The communal farming areas were deliberately set up as land, financial and natural resource poor areas. Successive governments deliberately produced and maintained communal area populations in a vulnerable state as a control tool. The nature of this state-induced vulnerability is gendered and has fostered dependency of women on men. Over time, the situation has been transformed and livelihoods have increasingly become vulnerable. Single women hearth-holds have become an important focus population as they are among the most vulnerable.

Communal areas were set up to depend on the hinterland consisting of urban areas, mines and commercial farms. Developments in Zimbabwe since independence and especially in the last ten years have resulted in the reduction of the migrant labour contract. This has had the effect of weakening the basis for economic linkages between communal areas and the rest of the economic space in the country. As a result, communal areas have increasingly become a disconnected and swelling reservoir of the unemployed, ill and redundant members of the population. Single women have been increasingly dependent on such communities for support. The declining livelihood base in communal areas means that such communities are increasingly finding it difficult to support single women and other dependent members of the community. Single women thus have to negotiate livelihoods in communities that offer limited support on difficult terms. The situation is dire for single women who have dependents, are ill and/or destitute. Single women have to devise strategies to diversify their livelihoods while coping with increased domestic burdens caused by HIV/AIDS, marital breakdown and the collapse of the public health system.

Communal areas have historically been perceived as areas where customary tenure is practised. The intervention by the colonial government disrupted and repeatedly re-organised natural resource governance institutions. This, coupled with changing land-use patterns as the viability of agriculture declines, means that the communal areas can hardly be said to have traditional or communal tenure regime. Single women may find new spaces and/or simultaneously lose security in the new and emergent tenure systems. It may be that traditional productive resource access has evolved in line with the new developments. The evolution may have opened up opportunities for marginalised groups. Examples include the informal land market and other institutions that accompany the evolution of resource governance institutions.

The vulnerability of single women in communal areas is rooted in the patriarchal structuring of communal area governance, which increased women's dependence on men. This patriarchal dominance has continued even as institutions for governing access to productive resources in communal areas have multiplied. Authority positions that were clear in the past have become blurred, and have caused vulnerability for women. Single women have moved from one position of dependence to a position of vulnerability in which they are less protected but have more possibilities to diversify on their own. The decisions that single women make determine the extent of their vulnerability.

The nature of rural-urban connections has been influenced by the economic crisis, government policy and declining agricultural production. The literature review has showed that rural and urban areas are inextricably linked. This has been illustrated by the knock-on effects that development in one locality have had on the other. The relaxation of laws on migration, land reform, fast track land reform and operation Restore Order have induced mobility of larger numbers of adults of both sexes to and from communal areas. The insecurity and uncertainty in commercial farms and urban areas have meant that people moving from communal areas make different decisions regarding their land rights in the communal areas. The rapidly changing rural-urban connections have induced changes in property relations and land use in communal areas in ways that have opened up land access opportunities for single women and other marginalised people. Single women have to make decisions to maximise opportunities and manage risks when they arise.

Although single women are the focus of this research, the impact of recent changes on communal areas and on the wider community in the research area forms an important backdrop against which to situate single women's experiences. Single women hearth-holds are an 'abnormal' category of domestic unit in an arena dominated by patriarchal norms. In the following chapters, the ways in which the communal area productive resources governance institutions have interacted with the increasing number of single women and the manner in which single women navigate, through their decisions, the volatility, uncertainty and desperation that characterise communal area livelihoods today will be discussed. The local context is an important determinant of the impact of processes outlined above. The next chapter focuses on the situation in the study area. This will provide a more specific backdrop for the empirical findings of the research, which will be presented in subsequent chapters.

Chapter 4
The profile of the research area

4.1 Introduction

The rapid changes experienced by the communal areas in the last two decades have increased livelihood differentiation and altered gendered relations. The impact of the rapid changes in each communal area was mediated by the location, resource endowment and local governance context. This chapter describes the specific local context of the study area. The focus is on two study villages. The aim is to highlight how different communities and governance institutions responded to the livelihood vulnerability caused by the crisis in Zimbabwe. The chapter is also used to provide a generalised framework within which the more focused analysis of single women hearth-holds in Chapters 5, 6, and 7 can be understood. The chapter is organised into four sections. The first section describes the background of Mwanza ward in which the two villages are located. It also describes the historic decline of the physical and social infrastructure available to the research populace. The second section highlights the governance institutions in the ward and the villages. The third section deals with the physical, social and governance aspects of the villages, specifically highlighting how individuals gained access to the villages and established homesteads. The final section describes the ways in which individuals dealt with vulnerability.

4.2 Ward governance

Map 2.1 shows the location of the research area in Mwanza ward, Goromonzi district in Mashonaland East province in Zimbabwe. Communal areas in Zimbabwe are governed by both administrative and traditional authorities who respectively have structures that are decentralised to the local level. Administratively, the villages are in Mashonaland East Province headed by the Provincial Governor. The Province is divided into Districts which are administered by the District administrator. Goromonzi district is divided into four wards. The ward, a sub-unit of the Rural District Council is headed by an elected councillor. The elected councillor is a member of the ruling party, ZANU-PF. A ward is made up of villages headed by a village head (*sabhuku*). The traditional authorities are represented by chiefs who have a council in the parliament of Zimbabwe. Map 2.1 highlights the communal areas in Goromonzi District. The Chiefs govern the communal areas in the district. The map shows that Goromonzi district is divided into the Chinyika, Chinhamora and Chikwaka Chieftainships. Chief Chikwaka's area is divided into three sub-chieftainships, whose spatial boundaries coincide with the administrative wards of the Rural District Council presided by the councillor. Sub-chieftainships are headed by sub-chiefs (*machinda*). The research villages are under the jurisdiction of Sub-chief Mwanza. Although the administrative and traditional authorities have set up parallel structures in communal areas, the ward and sub-chief's boundary coincide. This means the councillor and the sub-chief administer the same jurisdiction.[27]

The administration of both the traditional and administrative authorities' activities are both vested in the *sabhuku* at the village level. *Masabhuku* are the lowest tier of both the traditional and the political and administrative authorities. There are 35 villages in the study ward of Mwanza. The size of the villages ranges from 20 to 300 homesteads. The field research was carried out in Makuku and Ndamba village.

The political parties also assert their authority through both the traditional and administrative communal areas resource governance structures. At the time of the research, the ruling party ZANU-PF had a hierarchy of command that ran parallel to the governing system from the parliament right down to the village level in rural areas. In the rural district areas, the cell and branch of ZANU-PF were the same as the village and ward development committees respectively. In addition to the Zanu-PF ward councillor, the ruling party was represented by a ward co-ordinator at the ward level. The ward co-ordinator was supposed to be the ultimate authority in the ward, answering only to the provincial governor. The ruling party had an important role in mobilising the electorate during elections, governing the disbursement of government handouts such as seeds and fertiliser and the identification of target groups for donor-funded programmes. The war veterans who had settled on the surrounding former commercial farms controlled access to grazing, firewood, thatching grass, employment opportunities and other resources available on the former commercial farms. The politicisation of resources increased the vulnerability of communal area people by limiting their access to donor-funded resources and also increased interference with the targeting of state resources, as political rather than need-based criteria were used to select beneficiaries.

The chief and sub-chiefs administer justice in a number of domains such as domestic strife, conflicts between the communal area inhabitants, land allocation and adjudication of disputes between *masabhuku*. The local chief Chikwaka held a weekly court at the growth point on Sundays. The nature of the 'traditional' authorities varied in the research area. There was a village popularly known as UB40 in the

[27] Throughout the fieldwork, the tension between the elected councillor and the traditional authorities' representative the sub-chief was observed throughout the various public meetings. The tension arose because of a lack of clarity on which of the two was supposed to dominate the ward. I enquired into this and was advised that the sub-chief administers the traditional activities like rain making and *sabhuku*'s monthly meetings while the councillor presides over the political activities like rallies, and election campaigns and monthly ward meetings. However, in public meetings the two were constantly engaged in a power struggle. Activities like the distribution of Food Aid introduced new arenas for struggle. The sub-chief and the councillor tried to diffuse the tension by avoiding attending the same meeting together. When I enquired why they had not attended the meeting, they respectively claimed that they had not been aware of it or had other commitments.

Mwanza ward. This village was established by former commercial farm workers on the grazing land of Deera village during the fast track land reform programme in 2003. A former commercial farmer had approached a local *sabhuku* who was also a war veteran for land to establish a village for the former commercial farm workers. The *sabhuku* allocated them the village grazing land. The former commercial farm workers had demarcated their homesteads and had a former commercial farm worker as the *sabhuku*. Developments like this village introduced variations to the traditional norms and ways of governing land and livelihood capitals in the communal areas.

The Chikwaka spirit medium was responsible for all traditional ceremonies, including rainmaking, giving thanks to the spirits, healing and explaining inexplicable events like death, illness, barrenness and bad luck. The medium controlled resource access by regulating the exploitation of natural resources, which she deemed sacred. These included some plant species. She was also a source of non-agricultural employment for some people recruited to assist in beer brewing and certain rituals in exchange for food.

The local police office was located at the growth point. The police were responsible for law and order. Some criminalised activities, like beer brewing and vending, were important livelihood strategies for single women. There have been no veterinary officers in the village since the government stopped providing the subsidised dipping services. This increased the vulnerability of the communal area livestock to disease. Neither was the Natural Resources Board represented in the area during the fieldwork period. In the past, the Natural Resources Board had been responsible for evicting people who had settled illegally in the communal farming areas. The Natural Resources Board targeted people who settled along water courses, vleis[28], conservation areas and buffer zones between communal areas and commercial farms. In the past, non-governmental organisations had provided assistance to the community through funding and food transfers. The villagers mentioned that they had received food from World Vision, Help-Age and Care International. The NGOs had disbursed assistance to *masabhuku* whose responsibility was to identify target homesteads to benefit from the assistance. These activities had virtually ceased in 2005 because of the deteriorating relations between Zimbabwe and the West following the fast track land redistribution programme. During the fieldwork period, the government of Zimbabwe had provided every homestead in the ward with two kilogram's of beans for planting. The fact that every homestead had benefited could be linked to the forthcoming election (see Section 3.2.3).

[28] Wetland environments that retain water close to or at ground level throughout the year. They include marshes, sponges, swamps and reed beds (Bell and Hotchkiss, 1991:202).

4.3 Socio-economic description

Map 2.1 shows that Goromonzi rural district comprises communal, resettlement and commercial farming areas, and mines. The areas represent a cross section of livelihood opportunities. The research site is located in agricultural Region IIb (see Section 3.2). The area has a short rainy season lasting from October to February. The area's dependence on rain-fed agriculture renders the agricultural activities vulnerable to the frequent drought that characterises the country's climate.

The proximity of the area to commercial farms of Bromley and Arcturus made the research site a convenient destination for former commercial farm workers who had been evicted or retired from the farms. Although some of the former commercial farms were redistributed to war veterans under the fast track land resettlement programme, some European commercial farmers remained. These farmers and the war veterans were engaged in agricultural activities on some of the commercial farms. Commercial farming areas with their high labour demand were sources of temporary and permanent employment. Chikwaka communal areas' proximity to commercial farming areas, with their reservoirs of wage labourers made the communal area attractive for petty traders. The commercial farms were a source of stocks of produce[29] and commercial farm workers provided a market for women involved in trading activities.

Juru Growth Point (see Section 3.2), which is in Mwanza ward, is located on a busy highway linking Zimbabwe and the Southern African Development Community neighbours of Mozambique and Malawi. The highway is used by truckers moving within Southern Africa. The public services at the growth point include government offices, a post office, a bank, fuel stations, wholesalers, retail shops, a hotel, nightclubs and a clinic. The growth point acts as a magnet offering services to passing traffic and locals from surrounding farms and communal areas. There are two rural service centres in Mwanza from where locals obtain basic goods. There are several mines in the vicinity, which provide livelihood opportunities for small-scale miners. The Norah River which forms the eastern boundary of Makuku village is a source of alluvial gold.

The two villages and the district are linked to Harare by a tarred highway. During the research period, the road from the growth point to the villages was a dirt road, in poor state and served by private operators who used poorly maintained vehicles. The formal public transport providers cited low demand, political interference in pricing and routing, and the poor state of the roads as reasons for not servicing the

[29] Cabbages, tomatoes, beans, potatoes.

route. The road had been well maintained in the past by the government and former commercial farmers.

At the time of the research, there were two primary schools, two secondary schools and one clinic in Mwanza ward. The clinic was staffed by two full-time nurses. A provincial medical doctor visited the clinic once a month. The clinic also relied on voluntary staff drawn from the villages. There were fourteen village health workers in the ward. Twelve of these were women. In the past when the clinic had more resources, village health workers distributed food rations, some preventive medicines, contraceptives and gloves to the community. During the time of fieldwork in 2005, the clinic had no drugs. Patients had to supply their own stationery[30] and medication. The village health workers selected 'deserving' individuals to be 'seen'[31] by the doctor. This, plus the people's diminished income, reduced the numbers of people using the facilities and increased people's vulnerability to preventable and curable diseases.

The ward was also a beneficiary of the Zimbabwe Rural Electrification Programme, which was launched after independence. As a result, some wealthier homesteads were connected to the national grid and had electricity. There was an Anglican-church administered orphanage in the ward, which used to absorb orphans from the whole district. The orphanage was oversubscribed and had stopped accepting orphans. The other churches assisted some destitute members with food, medicine and moral support, and others purported to offer healing.

4.4 Village governance

The villages of Makuku and Ndamba were selected because of the striking differences between them. The differences are summarised in Table 4.1. Makuku was a much larger village in terms of population and size. The village had a total of 335 residents compared to Ndamba's 54. The average homestead size in Makuku was smaller than that in Ndamba because of the higher population density in Makuku village. Map 4.1 shows the layout of Makuku village. The village borders commercial farms to the east and, in the south, the boundary is a river. In the west, two other villages, Masawi and Tunha, are separated from Makuku by a river. The number of homesteads in Makuku village increased from 68 to 90 between September 2004 and November 2005. Map 4.1 shows that the homesteads in Makuku are divided into three clusters labelled Makuku 1, Makuku 2 and Makuku 3. The original settlement and officially established village was Makuku 1. This was later extended to Makuku 2 and Makuku

[30] Notebooks to be used by the clinic staff to record patients' medical condition and history.

[31] As there was only one doctor serving the whole province and there were no drugs, the doctor made monthly visits to monitor the sick. He could only examine the patients but had nothing to offer them by way of medicine.

Table 4.1. Summary comparison of Makuku and Ndamba villages.

	Makuku	Ndamba
Size (number of residents)	335	54
Number of homesteads		
2004	68	23
2005[a]	90	23
Homestead size	0.13-2.17 hectares	0.49-2.31 hectares
Occupied homesteads	68	15
Male-headed homesteads	39	9
Child-headed homesteads	1	0
Homesteads hosting single women hearth-holds	40	12
Women-headed homesteads	28	6
Membership access	Kinship, market, employment, marriage	Kinship, marriage
Social security	Kinship, market, state, church	Kinship, state, market
Livelihood opportunities	Agriculture, land market, commercial farming areas	Agriculture, commercial farms

[a]Homesteads depicted on the maps; number of homestead increased over the fieldwork period.
Source: Fieldwork 2004/5.

3. The resident population in Makuku consisted of 164 children, 89 women and 82 men in November 2005. Children made up nearly half of the population.

Map 4.2 shows the layout of Ndamba village, with Chipikiri village to the east, Goremusandu to the south and Mapfumo to the west. There were 23 homesteads in Ndamba in 2004. The number of homesteads in Ndamba village remained constant throughout the fieldwork period. As the map shows, homestead allotments were fewer and much larger than those of Makuku. The homestead sites in Ndamba ranged between 0.49 and 2.31 hectares in size. The population of Ndamba numbered 57 in September 2005, of which 23 women, 21 children and 10 adult men.

Villages are made up of individual portions called homesteads. Makuku had 68 homesteads in 2004 and Ndamba had 15 occupied homesteads. Makuku continued to grow because of the flourishing informal land market. This is the reason for the large difference in the number of homesteads between 2004 and 2005. The two villages proportionally had more male-headed than female-headed homesteads. The homesteads hosting single women hearth-holds were more than the homesteads headed by single women hearth-holds. Homesteads hosting single women refer to

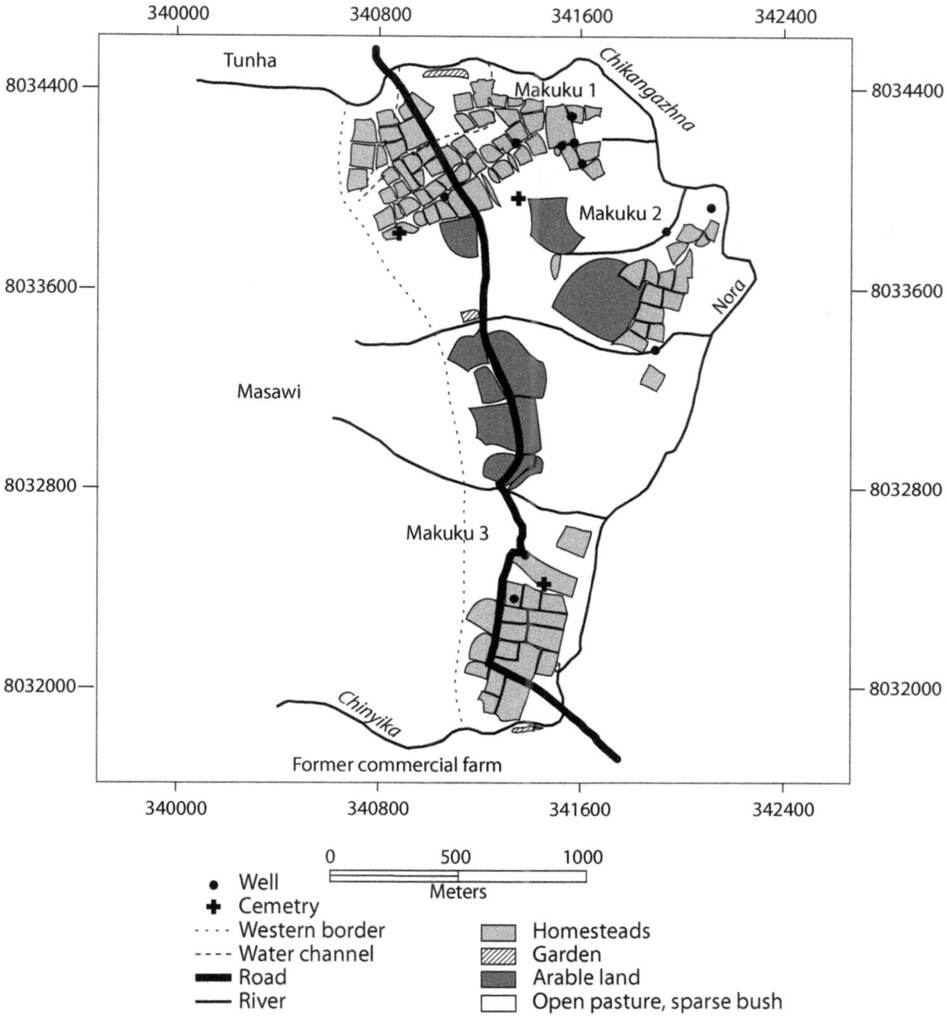

Map 4.1. Makuku village land use in September 2005. Source: Derived from GPS mapping procedure (see Sections 2.6.1 and 2.6.2).

homesteads where single women were accommodated both as homestead heads and dependents of either female or male head.

The village boundaries were widely contested. During the monthly ward *sabhuku* meetings and the Chief's Sunday courts, boundary disputes between *masabhuku* were commonly presented. Map 4.1 shows the boundaries of Makuku. The western boundary is marked by some very old trees. Land transactions also cut across village boundaries. Both Makuku and Ndamba villages had homesteads belonging to

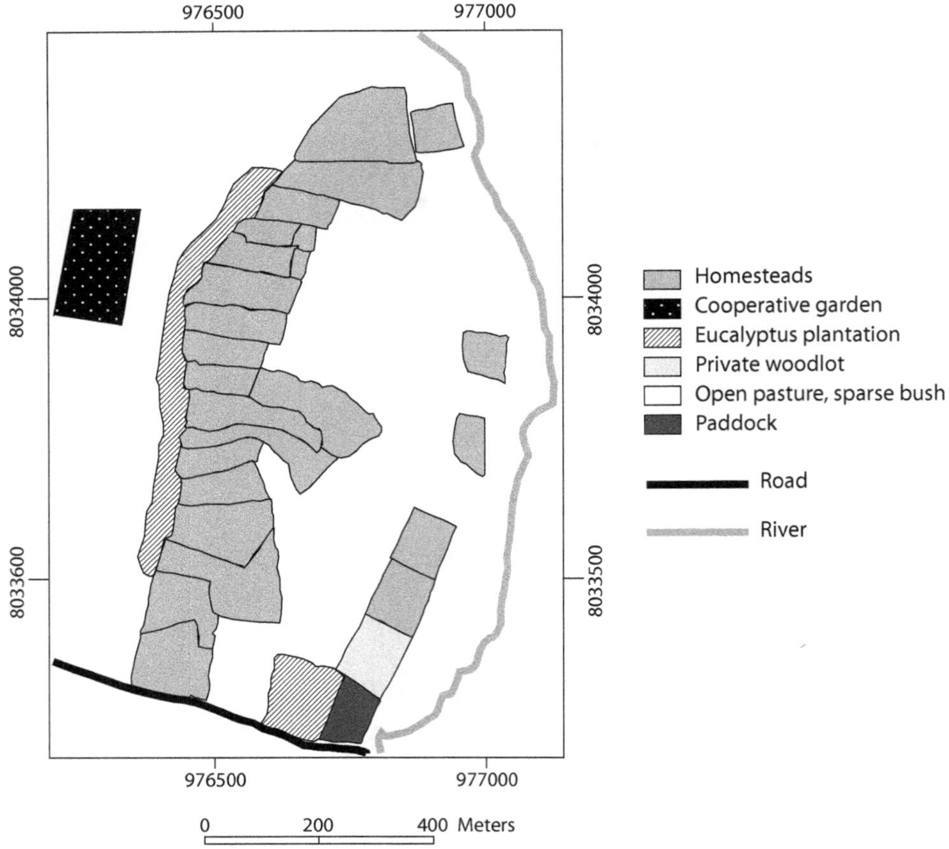

Map 4.2. Ndamba village land use in November 2005. Source: Derived from GPS mapping procedure (see Sections 2.6.1 and 2.6.2).

adjacent villages within their respective boundaries. On Map 4.1 depicting Makuku, the homesteads in Makuku of which three are located in the western part of the main road belong to Masawi village. The explanation is that the adjacent Masawi *sabhuku* had encroached on the territory of Makuku. Andersson's study in Buhera communal areas found that the practices of neighbouring villages encroaching boundaries were not uncommon (Andersson 2000:105). The Ndamba community garden on the western side on Map 4.2 is located outside its boundaries in the next village. Apparently, this is because of a transaction between the two *masabhuku* to facilitate the establishment of a garden[32]. The eastern boundaries of Ndamba enclose

[32] The site was suitable because it was in the vlei. Ndamba did not have such a site so the *sabhuku* exchanged a residential homestead for the garden site.

some gardens, which belong to homesteads from Chipikiri village across the river. These gardens existed before the establishment of Ndamba village.

In Ndamba, all homesteads had access to additional cropland beside the homestead. This land was used for growing crops. In Makuku, some of the homesteads did not have access to arable land because the land had been subdivided to accommodate the increasing population. Immigrants, younger families and the autochthons that sold and bought arable land were less likely to have arable land in Makuku village. There were people resident in Makuku whose livelihood portfolios were made up entirely of non-agricultural livelihoods. Such people did not have crop land.

Each village had access to a river from which they drew domestic water use and for livestock. Each village had originally been allocated some common land for fuel wood and grazing, usually vlei, rocky ground or other inhabitable areas. Settlement on these commons was a criminal offence that the State used to enforce through the destruction of the dwellings, imprisonment or payment of a fine. The institutions enforcing this were the Natural Resources Board and the police. Makuku had a larger grazing and forestry area because of its proximity to the commercial farming area. Originally, there was extra land between Makuku village and the former commercial farms. The extra land acted as a buffer between commercial farms and communal farming areas. The high migration into Makuku reduced the buffer, with the result that in Makuku homesteads bordered directly on the former commercial farms. The villages were served by service lanes, which were used for access by carts and cars, as well as for driving cattle to and from the grazing areas. Both villages had water channels that were low-lying areas demarcated by the government to facilitate drainage of rainwater and prevent flooding. Some individual homesteads had access to gardens located on vleis and riverbanks. The gardens were used for growing crops and vegetables. The produce was irrigated using buckets to carry water from the source which was usually a river or shallow well.

The Ndamba *sabhuku* was a retired labourer[33]. He had previously worked in Harare while his wife maintained the rural homestead. After retirement he became a full time resident in the village. The position of the *sabhuku* in Ndamba meant that there was a central authority around which the community was organised. The Ndamba *sabhuku's* wife led most of the donor-funded initiatives in the village. The Makuku *sabhuku* was a commuter. A tailor by profession, the *sabhuku* had a business in the town of Marondera approximately 60 kilometres away and another at a shopping centre in the adjacent village. The *sabhuku* commuted between the village and his business concerns. In his absence, his brother or cousin assumed leadership of the village but the villagers sometimes resorted to the chief or sub-chief. The *sabhuku's*

[33] Unskilled labourer poorly remunerated.

wife travelled with him. The two *masabhuku* differed in the way they governed their villages and received interpreted and practised state laws. The way in which this influenced livelihood decision making and resource access for the villagers is explored later.

4.4.1 Land access

People gained membership of the villages through kinship, the market, employment and self-allocation. The prevalence of various access modes varied between Ndamba and Makuku village. The residents in Ndamba were all related to the current *sabhuku*. They obtained land through kinship relations with the *sabhuku*. The only exceptions were an adult male who previously worked for the *sabhuku* and was allocated a homestead as a reward for his labour and two homesteads that belonged to former commercial farm workers who moved into the village at the height of the liberation war. The Ndamba *sabhuku* took some pro-active measures to deal with the increasing demand for land in the village. The aim was to suppress the demand for land by people not originating from the village and to secure land for the future generations of the village. If all land was allocated to the families currently in the village, the land stock would be exhausted and there would be no opportunity for outsiders and/or politicians to demand space in the village. At a village meeting in 2003, the *sabhuku* parcelled out the remaining village residential land in equal portions to all homesteads that were represented. Each homestead was allocated an extra homestead to cater for its children's needs. Any other children needing land would have to be accommodated from the individual family's allocation. The *sabhuku* restricted access to the village through various mechanisms, which are discussed in subsequent chapters. There were seven single women in Ndamba. Six of these hearth-holds headed their own homesteads. They gained access to the homesteads through marriage and natal claims. The homesteads were not originally allocated to the hearth-holds but as will be shown in Chapter 5, 6, they were allocated to other people before the women (husbands, parents). The women then continued to use these homesteads as widows, returned daughters or sisters of the original male to whom the homestead was allocated by the *sabhuku*. As a result, the homesteads that were occupied by hearth-holds at the time of the research fieldwork were not necessarily larger or smaller than male headed homesteads in the village.

Map 4.3 shows that Makuku village is demarcated into villages 1, 2 and 3. The situation in Makuku village reflected the wider diversity in the 68 homesteads. Twenty-four homesteads had obtained land from the *sabhuku* on the basis of their relationship with the agnatic core, that is, on marriage or upon attaining adult status. These were all settled in Makuku 1. The subsequent villages of Makuku 2 and Makuku 3 were established by the *sabhuku* in response to immigrant demands for land. Immigrants purchased land for homesteads on the informal land market. Map 4.3 shows that immigrants also settled in the original Makuku 1 village. The

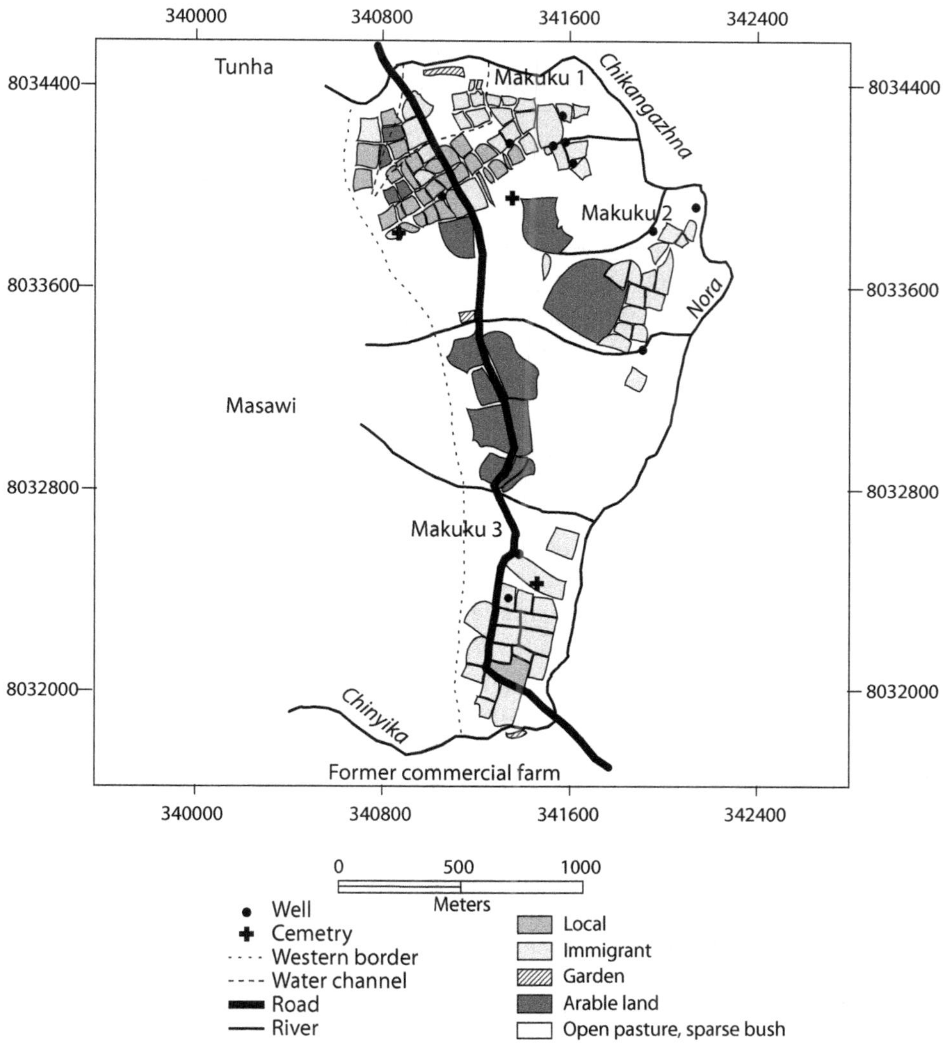

Map 4.3. Origins of Makuku village inhabitants. Source: Derived from GPS mapping procedure (see Sections 2.6.1 and 2.6.2).

immigrants in Makuku came from other communal areas, commercial farms and within Mwanza ward. The single local homestead on the map in Makuku 3 belonged to the *sabhuku* when he temporarily moved there. He has since disposed of the homestead on the informal market.

Nearly two-thirds of the people occupying homesteads in Makuku obtained their homestead from the informal market. They purchased land from the *sabhuku*, other villagers, other migrants or claimed land for themselves. The immigrants paid the *sabhuku* for the land and he 'paced'[34] out residential parcels. The immigrants then carved out gardens for themselves along the river banks.

Makuku 2 and Makuku 3 are located on Makuku grazing and forest. The land is rocky and impossible to plough with an ox- or tractor-drawn plough. Makuku 3 is inaccessible during the rainy season. The *sabhuku* assumed ownership of the communal commons as the customary trustee. He allocated plots in the former forest and grazing area to produce the settlements marked as Makuku 2 and Makuku 3 on the map. The act of parcelling out village commons to former commercial farm workers was also observed by Chimhowu and Woodhouse in another communal area in Zimbabwe (Chimhowu and Woodhouse, 2006b).

Research in Gokwe communal area in Zimbabwe reported that people in Gokwe sold their land because they were desperate for money and lacked the resources to use the land or to raise money for food during the 1992 drought (Nyambara, 2001a:266). Individual men and women in Makuku also sold their land.[35] For example: a male villager sold his arable land and used the proceeds to buy furniture. A widow (*sabhuku's* mother) sold some of the arable land that she inherited after the death of her husband. She used the money to invest in a pig-rearing project. Another widow also sold her arable land and used the money to fund the construction of a house. Autochthons (people born of the original Makuku agnatic core) subdivided their homesteads and disposed of the remainder. Some autochthons sold some or all of their arable land. Some immigrants who had bought into Makuku as compared to those originating from the agnatic core and or married to people related to the agnatic core group subdivided and sold land. Villagers sometimes sold their homestead plots when they left the village, whereas others retained control of their land. There did not seem to be a direct link between mortality and land sales in Makuku. Homesteads that had suffered mortality generally had the relatively largest land holdings, even though the land they held was generally under-utilised. The Makuku village secretary's minutes (Box 4.1) demonstrated that people gained access to land in the village through purchase on the informal market, self-allocation and allocation by the *sabhuku*. The minutes also showed that some people settled themselves in the village without approaching the *sabhuku*. The excerpts from the minutes of 28 December 2002, 27 March 2003 and 23 April 2003 all identified cases of people settling themselves without the knowledge of the *sabhuku*. When 'caught', such people were

[34] Originally, the state administrators, and later agricultural extension officers, measured parcels of land using surveying equipment and recorded the transactions. The apparent reference to the *sabhuku* 'pacing' land implies that he is not systematic and uses primitive and unscientific tools to undertake this highly technical and scientific task.
[35] It was not possible to get specific details of all monetary transactions involving land in the village because it is illegal to sell land in communal areas and the people involved in the transactions feared prosecution.

Box 4.1. Excerpts of minutes of Makuku village secretary.

28 December 2002
- complaints about people encroaching on the service lane, water channel, tree cutting
- *sabhuku* accused someone of making a garden without his knowledge
- someone settled themselves without knowledge of *sabhuku*

27 March 2003
- the village meeting people complained about inequality in the distribution of benefits of food aid
- the *sabhuku* said no squatters would be evicted if placed by his late father or those settled by him. Only those who settled themselves would be evicted
- someone had constructed a dwelling on the water channel and would be evicted
- all villagers had to pay money towards Unity Day celebration
- someone's cattle ate another's field produce; the owner of the livestock was fined two buckets of maize

28 March 2003
- the people present at the meeting complained that government aid beneficiary selection was done unfairly. They alleged that the neediest were marginalised
- cattle eating in people's gardens

23 April 2003
- three people settled themselves into the village without the knowledge of the *sabhuku*. They counter-accused him but he successfully evicted them
- the *sabhuku* reminded villagers that it is illegal to sell land and anyone caught should be reported to the committee

25 January 2004
- someone reported their caretaker for trying to take over their land
- someone settled without permission of the *sabhuku*
- illegal settler evicted
- *sabhuku* allowed some illegal people to settle

31 July 2005
- *sabhuku* selected food-for-work beneficiaries
- *sabhuku* selected people from the village to be recruited into the neighbourhood watch
- complaint from commercial farms about people poaching firewood

Source: Village records consulted during fieldwork, 2005.

fined and allowed to stay. Those who refused to pay the fine were evicted. An earlier section (4.4) of this chapter referred to contested boundaries. The adjacent *sabhuku* Masawi also sold Makuku land, claiming that forest and grazing were commons of both villages. Although the homesteads were spatially located in Makuku, they were administered by Masawi *sabhuku*. The various mechanisms through which people negotiate land access in Makuku show that traditional authorities are not necessarily the only institution responsible for land allocation in communal areas of Zimbabwe.

Immigrant headed homesteads made up seventy-six percent of the total homesteads in Makuku. A third of the immigrants moved to Makuku from the former commercial farms. The former commercial farm dwellers moved to Makuku on retirement or after the fast track land reform programme. A third of the immigrants moved into Makuku from other communal farming areas in Zimbabwe, to take advantage of the better agro-ecological conditions, escape land pressure in their own communal areas and to be nearer to Harare and opportunities for livelihood diversification. The rest of the immigrants moved to Makuku from the same communal area of Chikwaka but outside Mwanza ward. The main surges in migration into Makuku coincided with land pressure in other communal areas, the structural adjustment programmes, fast track land reform and operation Restore Order. Five new homesteads were carved out of the Makuku landscape to accommodate victims of operation Restore Order in 2005. These were all male-headed homesteads who came from Harare. Makuku 2 and Makuku 3 residents had been there for periods ranging from half a year to twelve years. The average period of residence of each immigrant homestead in Makuku 2 and Makuku 3 in 2004 was six years. People moving into Makuku from other communal areas in Zimbabwe were the dominant type coming into Makuku 15 years ago. Since then, people moving into Makuku from the commercial farming areas have been the majority of immigrants. This demonstrates the declining attraction of Makuku as a destination in comparison with other communal areas. This factor can be attributed to the increased price and/or difficulty of acquiring land and the negative effects of high population density.

Some immigrants initially arrived in Makuku as employees of the village inhabitants. These were employed to herd cattle, assist with farm labour or as house help. The number of employees increased during the agricultural season, though the majority ceased working after harvest. Some of these short-term employees married into the village and ended up residing there permanently.

Some people obtained land through a combination of the market and relations with the agnatic core in Makuku. The first group consisted of former commercial farm workers whose wives originated from Makuku. They negotiated access through their wives' relations and then purchased homesteads. The experiences in Makuku show that marriage was neither the only nor the dominant way in which people gained access to resources in communal areas. The market, employment, self-allocation[36] and politics provided arenas through which single women gained access

[36] Some people literally claimed land on the village commons for their private use. They fenced off portions of the grazing area and erected a residential structure. Others simply extended the size of the land that they had been allocated by the *sabhuku*.

to communal area resources. The decentralised land allocation in Makuku increased the opportunities for people to gain access to land. Ndamba offered fewer options for people to control land in its village. The Ndamba *sabhuku* suppressed market transactions by denying people who purchased land membership of the village and by confiscating land from Ndamba villagers who sold land. That such diversity can be observed in two villages in the same communal area makes it impossible to generalise about the effect of traditional and modern governance institutions or the informal land market.

Research in other communal areas of Zimbabwe has shown that an increase in population resulted in conflict (Andersson, 1999; Nyambara, 2001a:256). WLSA regional research concluded that increasing land pressure depleted woodlands. This increased the burden for women who are traditionally responsible for the provision of fuel wood (Mvududu and McFadden, 2001:98). The increased demand for land in Makuku resulted in some people encroaching onto water channels and public access roads. The excerpts from the Makuku village secretary's minutes illustrate some of the conflicts that have arisen between the different groups of people. Box 4.1 shows that conflicts were over land, firewood and the distribution of food-aid. The minutes of 31 July 2005 show that the conflict extended to commercial farming areas where Makuku residents collected firewood without the permission of the owners. The minutes of 27 and 28 March 2003 show that there have also been cases of land use conflicts between livestock rearing and cultivation. This was because the gardens were either not well fenced and/or were located on the cattle path (as was the case with gardens belonging to people who had bought into the village and had claimed land along the river banks for their vegetable gardens).

The Makuku community was divided over the issue of selling land, with those opposed pitted against those in support of the practice. The people who were opposed to the land market accused the immigrants of squatting. The Makuku residents opposed to the prevalence of the land market formed a squatter resistance committee. The local squatter resistance committee was led by the brother of the then *sabhuku* and supported by other people related to the agnatic core of Makuku village. The squatter resistance committee accused the immigrants of causing land shortage by increasing the population density, increasing crime as some of the immigrants were not employed, reducing grazing by settling on land that was apportioned as grazing land in the land-use plan, increasing river siltation by ploughing on the river banks and increasing the prevalence of corruption in the village by offering bribes to the *sabhuku*. The squatter resistance committee reported the matter to the Natural Resources Board. In 1999 the Natural Resources Board fined the immigrants and destroyed their gardens and dwellings. The 'squatters' reconstructed their dwellings. The *sabhuku* who had facilitated their access to the community endorsed this. Since the Natural Resources Board was no longer active in the district[37], the immigrants had established a permanent settlement.

4.5 Demographic analysis of the villages

The inventory confirmed the existence of a high number of single women hearth-holds in the two villages. In Ndamba there were nine male-headed, six female-headed and no child-headed homesteads. The number of homesteads hosting single women hearth-holds was 12 out of the 15 homesteads in Ndamba. In Makuku village, there were 39 male-headed homesteads, 28 female-headed and one child-headed homestead. Forty of the 68 homesteads in Makuku were hosting a single woman hearth-hold. Single women hearth-holds resided in various capacities as employees, divorced and/or widowed daughters and never married women. The phenomenon of child-headed households, which has been attributed to the incidence of HIV/AIDS, was observed on one homestead in Makuku but not in Ndamba village. In this case, the homestead was inherited from the children's deceased parents. Proportionally, Ndamba village had more single women hearth-holds than Makuku. This could be attributed to the continued existence of a relatively intact kinship-based social security system. The Ndamba *sabhuku*'s decision to suppress the land market in the village means that the Ndamba community was more cohesive than Makuku one. As a result, the land and community in Ndamba was better placed to offer traditional support to single women hearth-holds than could the more heterogeneous and less cohesive Makuku community. However, the more diverse ways of gaining access to land in Makuku opened opportunities for other categories of single women to gain access to communal areas. For example, single women entered Makuku through the market and employment, channels that did not exist in Ndamba.

The number of homestead residents on the occupied homesteads ranged from one to 12 in Makuku and one to 11 in Ndamba. Two homesteads were occupied by a single person in Makuku. One of these was a single woman. There were three homesteads occupied by a single person in Ndamba. All three were single women hearth-holds. The single person homesteads were more vulnerable as they lacked labour.

Homesteads in Mwanza lived with the burdens of suffering and care that come with very high morbidity. Thirty-one of the 68 homesteads in Makuku were hosting an ill adult. Twenty-two of the homesteads with ill residents in Makuku were single women hearth-holds. In Ndamba, six of the 15 homesteads were hosting an ill adult. The ill were equally represented in both sexes. The sick people were suffering from tuberculosis, respiratory problems, mental illness and afflictions related to old age.

[37] The Natural Resources Board was the government department responsible for upholding the conservation laws. In this case, they would be expected to mobilise the police to destroy any settlements that were seen to threaten the environment.

Four of the ill people in Ndamba were female hearth-hold heads. The other two were under the care of single women. Table 4.2 shows the number of ill people in Makuku and Ndamba. The table shows that HIV/AIDS-related illnesses were the most prevalent in the study villages. This increased the vulnerabilities of the ill and carers as it reduced time and labour available to maintain livelihoods. Age-related illnesses were also a source of vulnerability as the elderly were increasingly expected to take care of the 15-49 year-olds who were vulnerable to HIV/AIDS. The inability of the local clinic to provide health services increased the women's burden. The number of ill people indicates the risk of mortality faced by the study population.

The incidence of mortality since 1980 among homesteads was 11 out of 15 in Ndamba (Table 4.3).[38] The mortality figures show the full cost to human capital in the study population. In Makuku, 38 homesteads out of the 68 had suffered mortality. The table shows that HIV/AIDS-related illnesses were the main causes of death in Makuku homesteads and that the old-age-related were the lowest. The picture in Ndamba was similar. Table 4.3, though not drawn from a representative sample, gives an indicative view of the death rate in the study villages. The table shows that proportionally, the incident of people dying from old age related illnesses was very low. This means that the proportion of young adult deaths in the villages is very high. The table shows that the large proportion of deaths are due to opportunistic infections like TB, respiratory infections, pneumonia, meningitis and malaria. This indicates a high incidence of HIV/AIDS related mortality.

Table 4.2. Prevalence of morbidity at the homestead level in the study villages in 2005[1].

Village	TB/chest/pneumonia/ meningitis/malaria (AIDS-related)[2]	Old age-related	Backache	Mental illness
Makuku	15	3	11	3
Ndamba	4	2	1	1

[1] The researcher asked each homestead respondent to indicate the presence of sick people at the homestead in 2005.
[2] HIV/AIDS opportunistic illnesses. Illnesses that thrive on immune systems weakened by chronic infections (UNDP, 2004a:46).
Source: Fieldwork 2004/5.

[38] One of these homesteads had lost five adults between 1998 and 2000. The deceased included an elderly man and four adult daughters. All the deceased were either coughing and/or suffering from tuberculosis (HIV/AIDS related). All the daughters were single parents at the time of their deaths. Their children were taken care of by relatives.

Table 4.3. Number of individuals reported dead and the causes of death in the study villages since 1980.

Village	TB/chest/pneumonia/ meningitis/malaria (Aids-related)	Age-related	Other: (hypertension, poison, accident)	No idea
Makuku	46	2	10	6
Ndamba	11	3	4	1

Source: Fieldwork 2004/5

Eight out of 15 homesteads were fostering a child in Ndamba. Twenty five of the 68 homesteads in Makuku were fostering a child. The children were grandchildren, nieces/nephews or employees of the household head. There were no child employees in Ndamba. The grandchildren made up more than two thirds of the fostered children. The parents of the fostered children were either deceased or resident in urban areas, growth points or commercial farming areas. The two villages had a total of 147 children of school-going age[39]. Approximately ten percent of these children were not at school. This was because some guardians or parents lacked money to pay school fees, some children were ill, some children had been expelled from school and other children were employed. The employed children were all orphans.

The demographic profile shows that the majority of the resident population were dependents. The fact that these dependents were ill, aged and children indicates a shortage of a critical mass of human capital needed for agricultural activities. The morbidity and mortality data lie behind the increasing number and increased vulnerability of single women hearth-holds. This is because high morbidity and mortality allude to a loss of human capital and consequently hearth-hold labour and/or remittances. The high morbidity also represents an increase in the hearth-holds' burden of care. The large number of fostered children also indicates marital instability and an increase in the number of single women.

4.6 Living in the villages

A standard homestead comprises housing structures and a well for the provision of domestic water. While some households dwell together in a single homestead an individual that successfully establishes an independent homestead has more autonomy and in the case of hearth-holds, more control over their own labour. In

[39] Children in Zimbabwe attend school from the age of 6 until they are 19 years old.

order to establish a homestead, an individual or family first needs to secure access to the land. New homesteads depend on the availability of resources to cement relationships and meet the setting-up costs. Vijfhuizen's description of her experience in establishing a homestead in a communal area in Chipinge Zimbabwe emphasises the importance of financial capital, networks, natural capital in the construction process (Vijfhuizen, 2002:71, Andersson 2001: 100). The money is used to hire labour, purchase building materials and transport. The building materials consist of poles, grass and a door frame for a thatched hut. The house construction tasks are gendered. Men do the hard digging, cut the poles, lay the foundation and construct the roof. Women fetch water, gather stones, dig and transport sand, and plaster the walls and the floor. Women also collect the grass for thatching. Men are responsible for digging the well, constructing a toilet and the granary. If bricks are used, these are bought or moulded by both men and women using cement, water and river sand. Brick houses have more window and door frames, and asbestos sheets which cost money. Sometimes family members donate building materials. People use a combination of family and hired labour. The hired labour is paid in money. The labourers could also expect to be fed during the building process (Vijfhuizen, 2002:71).

The most secure shelter is that which is constructed using burnt bricks and mortar and roofed with asbestos and/or corrugated iron sheets. This shelter which requires less maintenance is relatively more expensive to establish. A survey in Murambinda communal lands in Zimbabwe in 1997 found that 84% of the homesteads that had a brick house were owned by returned migrant workers or their widows(Andersson 2001:100). This shows the importance of rural-urban connections in the establishment of homestead infrastructure in communal areas. The cheaper shelter is constructed from bricks that are not burnt. The roofing material is usually thatch and/or plastic sheeting. Structures made of pole and mud are vulnerable to the elements and while they require less capital in the outset, in the long run they cost more as they are vulnerable to wind, rain and termites.

Water is important for domestic use and gardening. People rely on shallow wells on their homesteads. People who have a well on their homestead also save time and labour, because they do not have to fetch water from a distance as do homesteads that do not have individual wells on the homesteads. Homesteads with individual wells are able to diversify into homestead based gardening, beer brewing and pottery activities. Well digging is a skilled task that is performed by men for a fee. The frequent droughts and fluctuating rainfall patterns mean that wells frequently dry up and have to be excavated. People, who do not have a well, have to collect water from other homesteads in the village.

A well developed homestead should have a toilet. This gives the people convenience and privacy. As the area does not have piped water people in the villages rely on pit latrines and ventilated improved toilets. Homesteads that do not have toilets have to rely on neighbours' toilets and or use the bush. An individual's capacity to mobilise the natural, social and financial capital to develop a homestead is illustrated by the type of infrastructure that they have on their homestead. Figure 4.1 shows the distribution of shelter, wells and toilets among individual homesteads in Ndamba village.

The figure shows that twelve of the fifteen homesteads that were occupied at the time of the fieldwork have brick under asbestos dwellings. All these homesteads had been developed with money earned by people who had been employed and or engaged in lucrative trading in urban areas. This illustrates the importance of rural urban connections for mobilising capital to invest in the development of the homestead. There were three homesteads that lacked decent housing. All three homesteads were run by single women hearth-holds. This shows the hearth-holds' relatively poor capacity to secure access to the money, networks and natural capital and their diminished rural-urban connections. Eleven of the fifteen homesteads in Ndamba had a well on the site. The proportion of hearth-holds that had a well was less than half of the total number of homesteads that had a well on the site in the village. Single women hearth-holds were the majority of the homesteads that did not have a well on site. This limited the women's capacity to participate in water intensive activities that were initiated by the various non-governmental organisations. The proportion of single women hearth-holds who had a toilet in Ndamba was less than half of the total number of homesteads that had a toilet in the village. The overall trend for Ndamba village shows that while twelve homesteads have decent housing, only eleven of them have a well and nine have a toilet on their homesteads. This shows the limited capacities of homesteads to either establish and/or maintain the non-housing infrastructure on their homesteads.

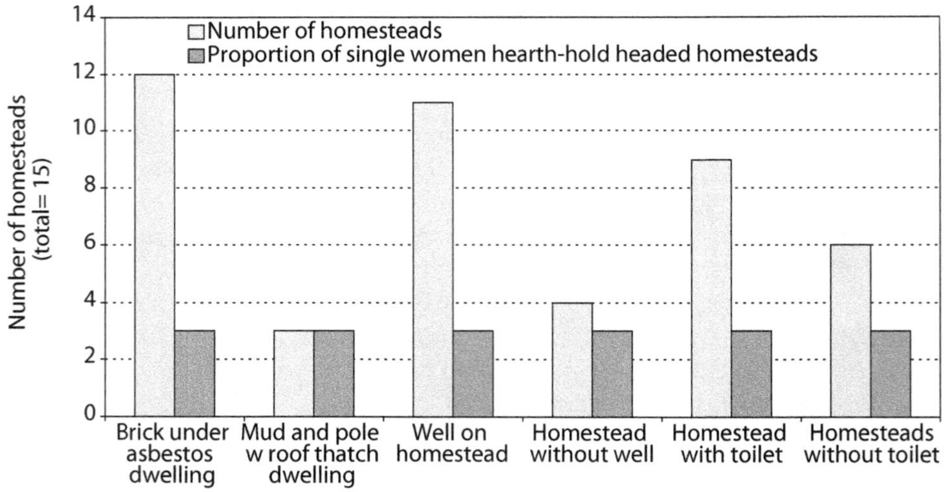

Figure 4.1. Homestead access to infrastructure in Ndamba village in 2004.

Figure 4.2 shows Makuku homesteads access to infrastructure. The figure shows that 40 of the 68 occupied homesteads in Makuku had a brick and asbestos structure. Single women headed hearth-holds made up less than half of the homesteads with access to decent accommodation in the village. However, the proportion of homesteads without access to decent shelter made up nearly half of the 68 homesteads. Twenty seven homesteads in Makuku did not have a brick and asbestos housing dwelling. Some of these homesteads had structures that had been abandoned during construction because of lack of funding and building materials. Sixteen of the twenty-seven homesteads that had substandard shelter were headed by single women hearth-holds. This over representation of single women highlights their limited capacity to construct adequate shelter. Not all homesteads that had a brick and asbestos structure had a well in 2004. The number of single women hearth-hold homesteads that had a well was 11 of 38. This shows the extent of difficulty that not only single women but also male headed homesteads had to excavate and maintain a domestic water source in Makuku. The high proportion of homesteads without a well in the village is also due to the fact that all of Makuku 3 village relied on a single borehole for water. This was because of the low water table in Makuku 3. The borehole was drilled by a person who had since abandoned his homestead in the village. The other homesteads without domestic water relied on the river, gardens and neighbours for domestic water. This water was availed at no charge. The number of homesteads that did not have a toilet outnumbered those that did in Makuku. This is partially due to the relative poverty of the former commercial farm workers who lived in the village and their limited capacity to mobilise financial and social capital to invest on the homesteads. The proportion of homesteads without toilet in Makuku was higher than the number of homesteads that had a toilet on site. This again is due to

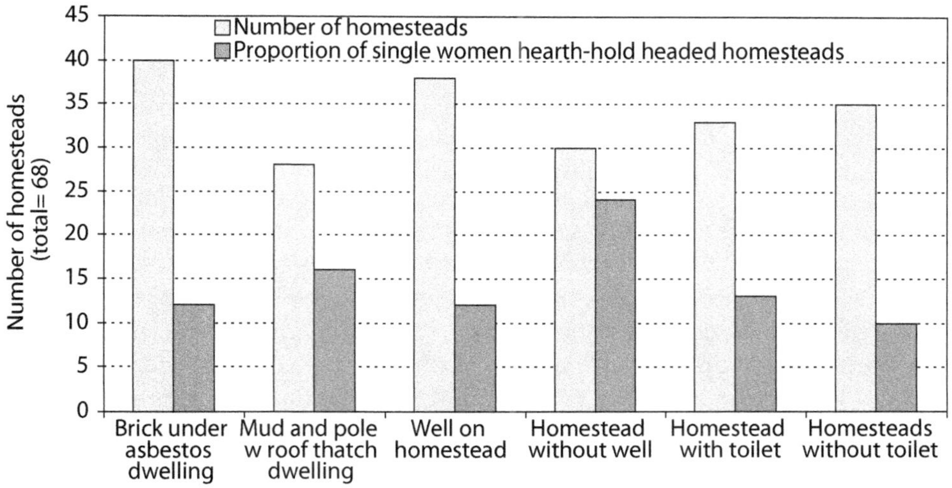

Figure 4.2. Homestead access to infrastructure in Makuku village in 2004.

the relative poverty of the population, shorter period of tenure and the existence of a significant proportion of former farm workers among the population. A focus group discussion with the *sabhuku* revealed that people financed the construction of structures on their homesteads with money from crop sales and or proceeds from wage labour. The decline in people's capacity to mobilise financial capital from farming and or wage employment is reflected in the declining housing and homestead infrastructure standards.

The rest of this thesis focuses on single women hearth-holds. However, it is important to recognise that the difficulties that the single women hearth-holds confront are not restricted to single women. The following cases, the first a young couple and the second an older man, illustrate the social construction of residential rights and livelihoods in the context of rural and urban economic and political turbulence in Zimbabwe. The examples draw on experiences of male headed homesteads in order to place single women's experiences in the perspective of other people's experiences.

4.6.1 Establishing a communal area homestead

This young couple married during the fieldwork period. The man, aged 23 years, was living in Harare and doing informal carpentry work. The man rented some accommodation while he was employed and returned to the communal areas when he was not working. His wife resided in the village with her in-laws. After the birth of their son, the couple were allocated a piece of land from the land that had belonged to the man's deceased grandmother. The man's family assisted the couple with labour for land clearance. The couple used the husband's wages to pay people to mould bricks. The couple constructed a kitchen using money earned by the husband from his work. The wife and son moved into the house. After six months, they constructed a two-bedroom brick house structure and hired someone to sink a well. The man's parents gave them a cow to start their herd. They also bought a pig. The man who had worked as a carpenter in the informal sector in Harare lost his employment during operation Restore Order. He returned to his rural homestead where he resumed full-time residence. After three months, the Ndamba *sabhuku's* wife helped the man to secure a contract to repair furniture at the local schools and clinic. The man's wife focused on subsistence farming and trading commodities from her homestead. The case shows how nuclear households gained access to land and established a homestead in Ndamba. In addition the case illustrates the importance of social relations in securing livestock, land, labour and providing assistance when the household was vulnerable. The networks helped the family to access the 'seed' livestock, labour and access to employment. The rural-urban connections were important for mobilising financial capital to fund the homestead construction. However, the delicate connection between rural and urban areas was managed by the family maintaining two bases to guard against adverse developments in either sector. The utility of the rural area homestead as a fall-back position when the man lost his urban employment illustrated the reason people choose to maintain their rural land claims in Zimbabwe.

In addition to the former commercial farm workers, retrenched people, Restore-Order victims and men in both villages, including those currently employed, failed to establish independent homesteads. These continued to live with their spouses in their parents' homesteads with their parents and/or siblings. In both Makuku and Ndamba, there were married men who had been unable to develop the homesteads that they had been allocated. There were four such men in Ndamba. The number of such men in Makuku could not be established accurately as some of them acquired their land on the informal land market. Some men moved back into their parental homesteads after suffering retrenchment, death of a spouse or illness. The man cited in the young couple case managed to survive the consequences of operation Restore Order by drawing on his kin and returning to the communal areas. A young man in Ndamba sold his land to obtain cash, but the *sabhuku* repossessed the land and evicted the buyer. There were also men who lived in both Makuku and Ndamba but lacked resources to develop their own homesteads. They managed by living on land and dwellings belonging to migrants. The following case describes this.

4.6.2 Securing a place to live

The old man was not sure of his age, though his identity document showed him to be 94 years old[40]. He had come from Mozambique when he was young. He left a wife and son in Mozambique. He worked on a commercial citrus farm all his life. He left after the farm was repossessed under the fast track land reform in 2001. He had no known relatives. He lived in Makuku 3. The homestead was not cleared but was well fenced. The homestead belonged to a couple who were living and residing on a commercial farm in the surrounding area. The couple who bought the place constructed a grass thatched hut and a two-bedroom brick and asbestos structure. There were gaping holes where the door and window frames would be fitted. The old man put some old sacks to protect the doorway. He guarded the homestead for the couple in return for residence. He worked on other people's homesteads in Makuku to earn his food. This was one of the most vulnerable cases in the village. On being asked what would happen if he fell ill or died, the old man replied: *'Just throw my clothes into the grave with me, then that will be that...'*

[40] It is common for elderly Africans in Zimbabwe not to be aware of their birth dates. However, such people usually obtained identity documents to enable them to work in former European areas. It is not uncommon for the age records on such identity documents to have no relevance to the owner's actual age.

The case shows that both single men and women are vulnerable. The man lost his 'pension' when he was evicted from the commercial farm where he had worked[41]. The old man's access to the homestead was assured as long as the couple who purchased the homestead continued to reside at the commercial farm. The couple who purchased the land use it as security and a fall-back position in the face of insecure tenure they have as commercial farm employees. The couple's absence provides an opportunity for the man to have access to residence and community membership. At his age, there was very little the old man could do to secure his livelihood. The old man's vulnerability was heightened by his immigrant status which limited the social support he could mobilise from the village where he resided.

Some people resided in Makuku because of the opportunity that the village offered them to pan for gold. These people constructed temporary structures which they used for the 3-4 months of the year when they were panning gold. The gold panners' interest in the village was limited to their ability to extract gold from the river at a certain time of the year. As a result, they were not in a position to invest in the development of a homestead, acquisition of agricultural land or accumulation of livestock in the village.

4.6.3 Agricultural practices

Agricultural activities formed a substantial part of homesteads' livelihood portfolios as they enabled people to produce food for their own consumption and possibly raise cash to finance other activities from the sale of their own produce. The agricultural season lasted from October to June. The village inhabitants farmed on the homesteads and on their arable land units. In order to engage in agricultural activities, a homestead needed to have access to arable land, plough, livestock, labour and seeds and fertiliser. These were procured using the money earned through rural-urban connections. Livestock has traditionally been used as a way to accumulate savings and disposed of to raise money during times of crisis (Bell and Hotchkiss, 1991:208; Scoones *et al.*, 1996). At the time of the research, cattle ownership was low and declining in the two villages because of drought, disease, the increasing incidence of livestock theft and limited cash flows on the rural-urban continuum. The main cause of draught animal loss was red water disease. The total number of cattle in the two villages was 128 in November 2005. Sixty-one of these cattle belonged to Ndamba village. Cattle ownership in Makuku village was low as 77% of the people in the village did not own any livestock. The average cattle ownership in Ndamba was 4 cattle per homestead while that in Makuku was less than one head of cattle per homestead. The differences can partially be explained by the high population density and mobility in Makuku and the prevalence of former commercial farm workers in

[41] The man had planned to see the rest of his days out at the commercial farm where he continued to have access to housing and subsidised food and had developed a network of friends over the years.

Makuku who were a relatively poor community. The low incidence of cattle among Makuku homesteads is also partially due to the reduced grazing and more dense settlement of the community which leaves little room for agricultural activity and livestock grazing. Earlier in the chapter, it was pointed out that agriculture is not a portfolio activity for some of the Makuku homesteads especially those who obtained village membership to enable them to pan for gold in the river or to be nearer to their place of work in Harare. This also partially explains the low incidence of cattle ownership in Makuku. Forty-six of the homesteads in Makuku had never owned any cattle. Two of the homesteads that owned cattle in Makuku were headed by single women hearth-holds. The single women hearth-holds had obtained these livestock through inheritance and on the marriages of their daughters. Single women were under-represented among the homesteads that owned cattle in Makuku.

In Ndamba, 11 homesteads out of the 15 owned cattle. Three of the homesteads were single women hearth-hold headed homesteads. Three out of the four homesteads that did not own cattle in Ndamba were run by single women headed hearth-holds. The low incidence of cattle ownership among the homesteads reflects a lack of savings among the households. This also shows the limited capacity of the village to engage in agricultural production as cattle are also used for land preparations. The low cattle ownership also reflects the limited capacity of communal area populations to save.

Sixteen homesteads in Makuku owned ploughs. Two of the homesteads that owned ploughs were single women hearth-hold headed. The homesteads that had ploughs bought them when the Zimbabwean economy was relatively robust. Others had inherited them upon the death of the people who had purchased them. Fifty-two homesteads did not own ploughs. Twenty-four of the 52 homesteads that did not own ploughs were headed by single women hearth-holds. The figure also shows that some homesteads in Makuku owned ploughs but had neither cattle nor arable land to plough. These were male headed homesteads that had disposed of their arable land on the informal land market. In Ndamba however, 11 of the fifteen homesteads owned ploughs. The majority of the households in Ndamba had inherited their ploughs. The capacity of the homesteads in both villages to secure and maintain agricultural assets is linked to the changing rural-urban connection sin Zimbabwe. During the interviews, respondents who had previously owned cattle had failed to replace them after a loss to drought or disease. This mirrors the limited decline in urban to rural cash flows which had historically been used to accumulate agricultural equipment and inputs in communal areas.

The villages were originally set up in such a way that each homestead was allocated a garden in the vlei and arable land for agricultural production. The Figures 4.2 and 4.3 clearly illustrate that while homesteads have access to arable land in Ndamba village, in Makuku, more than two thirds of the sixty eight homesteads did not have arable land. More than twenty of the homesteads without arable land were headed

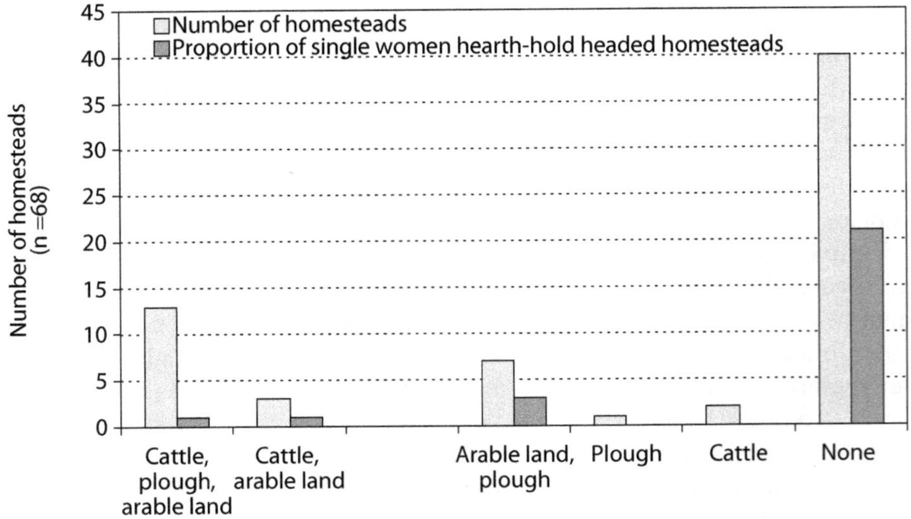

Figure 4.3. Makuku village homestead access to agricultural assets in 2004.

by single women hearth-holds. In Makuku, the homesteads without arable land relied on the homestead plot for their farming activities or entered into various arrangements with those who had a surplus of land and/or could not utilise all of their land because of other limitations for arable land. There were 22 single women hearth-hold headed homesteads in this category. There are three main reasons for this pattern. The land market reduced people's access to arable land. Some people resided in Makuku village solely for the opportunity the location offered them to exploit the growth point, urban areas, mines and former commercial farming areas. They either lacked the capacity to undertake, or had no interest in, subsistence agriculture. The other immigrants who came to the village in search of the alluvial gold panning opportunity only resided in the village during the gold panning season and emigrated for the rest of the year. Such people did not acquire communal area land for agricultural purposes. The elderly retired former farm workers were too old to engage in any meaningful agriculture and struggled even to farm on the small piece that they had. None of the people who had arable land in either village utilised all of their land, though they claimed to have previously used all the land in the past. In Ndamba every resident in the village had access to arable land for agricultural use. However, the homesteads had a reduced capacity to work the resource due to shortages of labour and inputs. The only time anybody in Ndamba had used their entire arable field was in 1982 when the government provided all the inputs. Since then, an increasing amount of land has been left fallow.

Thirteen homesteads in Makuku had access to arable land, livestock and a plough. All except one were male headed homesteads. There was only one single woman headed hearth-hold in this category. Figure 4.3 shows that other homesteads in Makuku possessed two or one of the three basic assets needed to engage in agriculture. More than two thirds of the homesteads in Makuku did not possess land, livestock or a plough. This illustrates the limited capacity of the village to engage in agricultural production. Even if one considers the various transactions that households enter into to increase access to agricultural implements, the capacity for villagers to engage in subsistence agriculture is limited. Proportionally, single women headed homesteads lag behind male headed homesteads in their ownership of agricultural assets.

The rest with neither arable land, ploughs nor cattle, resorted to various networks which allowed them to hire draught in exchange for money or services, or to obtain draught free of charge, perhaps from relatives. The homesteads that could not mobilise draught resorted to hand-hoeing their fields. The terrain in Makuku 3 made it impossible for villagers to use ploughs or tractors. The homesteads suffered labour shortage because of migration, illness and the unwillingness of the young people to engage in subsistence farming. More than two thirds of the homesteads in both villages relied on their own labour for subsistence farming. Less than a third hired labour. Figure 4.4 shows the distribution of agricultural assets in Ndamba village.

All homesteads in the two villages benefited from the government seed transfers for the 2004/5 season. However, the government seed was delivered late. The late arrival of government seed and the short planting season increased the risk of crop failure for the households who were unable to arrange alternative sources of seed. Some people planted the seeds but did not harvest anything. The rest either used seed from previous harvests or purchased the seed from the market. In Makuku, 46 homesteads purchased agricultural inputs. Twelve of these were single women. Nineteen homesteads used previous years' seed; eight of these were single women.

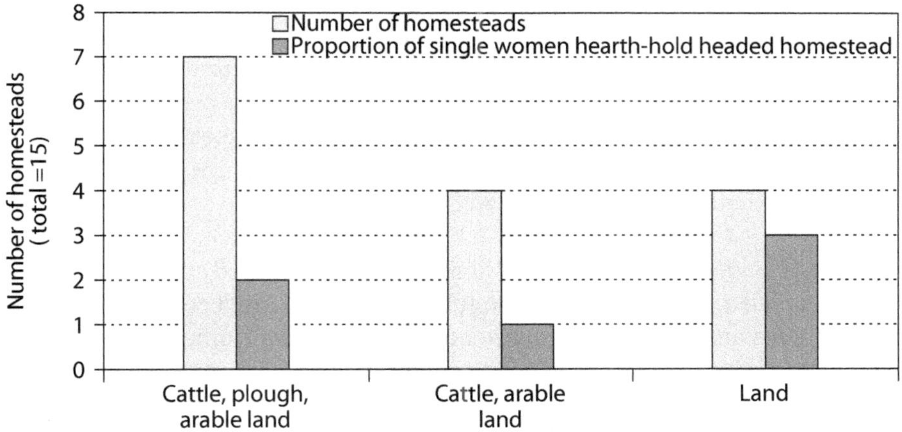

Figure 4.4. Ndamba village homestead agricultural assets access in 2004. Source: Fieldwork 2004/5.

People who use old seed run an increased risk of crop failure because old seed loses its potency (Folta and Deck, 1987:338). In Ndamba, four homesteads (two single women and two married couples) bought seed. One homestead obtained seed from the husband's employer. Two single women got a loan from the dairy project (see Section 4.7, NORAD heifer project). Seven single women relied solely on the government for agricultural inputs.

There is also a clear relationship between the homesteads that can mobilise agricultural inputs and the extent of their rural-urban connections. The changes in rural-urban connections induced by the deteriorating economic and governance condition in the country are reflected in the changing livelihood portfolios because increasingly agricultural oriented portfolios centred around the investments in land and agricultural implements like ploughs and livestock funded by urban remittances. The homesteads headed by younger men and women and former commercial farm workers had the most difficult task to mobilise agricultural inputs. These homesteads typically had limited access to urban cash flows that were secured through employment and remittances. The limited capacity of communal area residents to produce their own food means that they had a diminished capacity to subsidise urban livelihoods through transfer of food.

Table 4.4 shows that agricultural production was becoming an increasingly risky activity in both villages. None of the households marketed any of their harvest[42].The homesteads that harvested more than their subsistence needs elected to retain their surplus to offset future food shortage. Households defined as 'too poor to farm' are those that lack the resources to buy, or creditworthiness to borrow, minimal inputs or to get labour and continue to farm by hiring themselves out in exchange for the commodities they lack (Whitehead 2004: 2). Sixteen percent of the homesteads offered their labour to work for those who had adequate inputs. They assisted with weeding, harvesting and processing maize in exchange for food and/or cash. Both elderly and younger men and women employed this strategy. Some of the land rich but input poor elderly leased out some of their agricultural land to land-short households and benefited from the harvest of their tenants. Single women were involved in these transactions, as ten of them controlled excess arable land. This arrangement was widespread in Makuku where there was a wide discrepancy in land ownership between the homesteads.

Vlei gardens provided an alternative for homesteads with limited access to resources to engage in rain fed agriculture to produce subsistence food. Produce from such gardens usually includes an early crop of maize and vegetables for home consumption and there may be surplus for sale (Bell and Hotchkiss, 1991). In Ndamba, every homestead

[42] The annual Food and Early Warning System Project used a consumption threshold based on the Zimbabwe national grain consumption which indicated 250kgs of maize per capita (Dekker, 2004:185).

Table 4.4. Maize harvest (per homestead) for the 2004/5 season in the study villages.

Village	Number of 90 kilogramme bags of shelled maize harvested			
	>10	5-10	<5	None
Makuku		44	13	11
Number of single women hearth-holds		23	10	5
Ndamba	3	8	2	2
Number of single women hearth-holds	1	6	2	2
Total	4	81	27	22

Source: Fieldwork 2004/5.

had access to a vlei garden plot. In Makuku, 35 of the 68 homesteads had access to a vlei garden. However, in both villages, the vlei gardens were under-utilised because the gardens were located away from the homestead, making them vulnerable to theft, domestic livestock and wildlife. The constraints reduced women's ability to benefit from them. Garden productivity was also constrained by the availability of labour and water. Water was abundant during the rainy season and dried up thereafter. Labour demand for subsistence agriculture was highest in the rainy season.

Some women coped by concentrating on the kitchen gardens which were located on the homestead plot. These gardens are small fenced plots used for growing vegetables. Since gardens have to be watered from domestic wells and kitchen wastewater (Bell and Hotchkiss, 1991:210), people without their own source of water, or whose fence is ill-maintained, cannot have a kitchen garden. Unfenced gardens are vulnerable to goats and cattle. The immigrants in Makuku occupied gardens along river banks. This brought them into conflict with the Makuku residents who have historically been using these places for grazing livestock. The conflicts were recorded in the excerpts from the Makuku secretary's minutes. The Ndamba villagers reduced the vulnerability of individual gardens by establishing a community garden. This is discussed in Chapter 7.

The women in Ndamba were engaged in gardening, subsistence agriculture, non-governmental initiatives, and one was employed as a local house worker. The women in Makuku were involved in subsistence farming, trading or *maricho*. There were also women who worked on commercial farms, were employed as maids in town or engaged in beer brewing. Some women either looked after the ill and/or were too ill to work and remained dependent on other people. The limited capacity for agricultural production meant that homesteads had to diversify into non-agricultural activities to sustain their livelihoods.

4.6.4 Other livelihood options

The people had to diversify to sustain their livelihoods. Although altered significantly, rural-urban connections were important in homesteads' livelihood construction. The breakdown on the migrant labour connection resulted in people struggling to develop homesteads and maintain an agricultural centred livelihood. An examination of the occupations of men in the two villages illustrated how individuals struggled to manage the volatile conditions. In Makuku 34 men were resident in the village at the time of the research. Two of them were employed as teachers in local schools. Even then one of them supplemented his low salary with running a tuck-shop. This man lived on a well developed homestead that he had inherited from his father who had built the main house. The other teacher had failed to construct his own homestead and relied on his widowed mother's housing for his own accommodation. The other men earned salaries as general cleaner at school and a driver on the commercial farms. These men's salaries were inadequate to develop their homesteads. Their wives and children lived as subunits on homesteads belonging to the men's parents while their own homestead remained undeveloped. Eleven of the men listed as their main trade, building, brick moulding and farm work. The rest of the men combined gold panning, poaching, and golf caddying and informal trading. These occupations reflected the men's attempt to adjust to the contracting opportunities for wage employment in urban areas.

The men in Ndamba relied on cattle herding, pensions, building and temporary work that involved well digging, work on commercial farms and piecework. The opportunities for these activities were curtailed by the declining capacity of people to invest in building in communal areas itself a consequence of the rapid contraction in urban wage employment opportunities.

Mobility was, and still is, an important strategy for mitigating vulnerability. Men from the study moved between the rural and urban areas to secure livelihood capitals. The absent men were reported to be engaged in low paying work such as caddying at golf courses, driving tractors on commercial farms, security guard and general hand. The women worked as domestic workers or shop keepers at the growth point, or engaged in trading. The exceptions were women and men who travelled to South Africa to trade goods. The marginal occupations realised insufficient income to enable the people to invest in and maintain infrastructure in the communal areas.

Some men and women were unable to travel out of the communal farm area because of illness, lack of opportunity, advanced age or the need to take care of care of small children. These people engaged in local wage labour for other residents and the schools, dairy, orphanage and the clinic. This included cultivating, collecting firewood, housekeeping, cattle herding. The homesteads that provided such opportunities had access to a reasonably steady source of income to pay for these services. These opportunities were limited by the declining liquidity of communal area residents which in itself was a consequence of the contraction of the urban economy of Zimbabwe.

Thirty-six of the sixty-eight homesteads in Makuku reported receiving remittances. Eighteen of these were headed by single women hearth-holds. The women's children sent the remittances. Thirty-two homesteads did not receive remittances. Nineteen of these were headed by single women hearth-holds. Twenty of the homesteads that did not receive remittances used to receive remittances in the past but these stopped following the death or retrenchment of the remitter. Fourteen of these homesteads were headed by single women hearth-holds. Eight homesteads had never exported labour and therefore never received remittances. Five homesteads had access to an urban immovable property. These belonged to elderly widows and retired men in the village. These properties were potential sources of rental income for the owners who lived in the village. Half of the people who owned urban immovable property received rents. The rest did not receive any rent because these properties were occupied by other members of the homestead, for example children of the owners, who did not pay any rent.

There were three homesteads receiving remittances in Ndamba. Two were married women whose spouses were working in urban areas. One was a widow receiving support from her adult children. Twelve homesteads did not receive remittances; nine of these were single women. Two male-headed homesteads were receiving pension during the fieldwork but this was eroded by inflation. Three homesteads owned immovable property in urban areas and received rent from their properties. The inventory also highlights the fact that 82% of the homesteads had experienced theft of items including livestock, household items, clothing, food and agricultural equipment.

Bebbington identified rural proletarisation as a significant growth of a rural proletariat working on capitalist agricultural enterprises. Although wages are low and hazards are high, this can at times resolve the rural residence/making a living dilemma by enabling people to remain in their communities by earning income from labouring (Bebbington 1999). As already stated, in Zimbabwe, commercial farming activities have declined following the fast track land redistribution. This has negatively affected industrial growth in the country, which is largely agro-based. A study by Adams (1991b) in south-eastern Zimbabwe revealed that single women dominated casual farm labour. Commercial farm work is a highly stigmatised activity among the indigenous Zimbabweans. Men are reluctant to do it and discourage their wives as they feel it reflects a man's failure to cater for his family's needs. The poor working conditions also discourage indigenous Zimbabweans (Chadya and Mayavo, 2002:13).

4.7 Community organisation

Community resources include moral support, obligations, assistance, co-management of community resources, security, a sense of belonging and assistance with children. Ndamba and Makuku villages exhibited great differences in the nature of community support and obligations to their members.

The limited influx of immigrants into Ndamba has ensured the relative cohesiveness of the community when compared to the situation in Makuku village. The nature of the relationships in Ndamba village enabled the villagers to organise collective action and assist each other. When a homestead suffered mortality or had an ill member, the others took turns to visit and assist the afflicted homestead with food, money and moral support. Generally, people went beyond the obligatory assistance. The extra resources were also mobilised through the kinship network and the centralised control of village governance by the *sabhuku*. The wealthier households in Ndamba subsidised the poorer ones. Three single women were too poor to make the monthly funeral insurance contribution but they still received assistance when their homesteads suffered mortality. The funeral contribution record showed that absent people who considered themselves members of the village also contributed towards other funerals in the village. The cohesion and co-operation among Ndamba community has also enabled it to benefit from donor-initiated projects.

A non-governmental organisation called Fambidzanai Trust encouraged organic maize farming and herb gardening. The trust provided herb seedlings to participants. Participants were encouraged to use domestic wastewater for kitchen herb and vegetable gardens. The produce from the gardens improved nutrition and was a possible source of income as there was commercial demand for the herbs[43]. All homesteads in Ndamba with the exception of two widows participated in the initiative. Fambidzanai Trust project also established kitchen gardens. The two women who did not participate did not have a well on their homestead. The community shared seeds and innovations that they received from the non governmental organisations.

NORAD initiated a heifer project in Goromonzi District in 1984 in collaboration with the Zimbabwe government, former commercial farmers and Heifer International. The aim was to restock the herds of farmers in communal areas, provide members with livestock and consequently milk and manure. The project targeted the poor members of the community. The Government of Zimbabwe and Heifer International

[43] At the time of the study, the Rural District Council was buying herbs from the community to package for commercial sale.

purchased livestock from the commercial farmers. The livestock were passed on to communal farmers after they had fulfilled certain conditions: the farmers had to construct a shed and a paddock on their homestead after which they were allocated a heifer[44]. A focus group discussion with the members of the co-operative revealed the paradox of the project: the project aimed at increasing the communal area head. While targeting the poorest in the community, the type of livestock disbursed by the scheme was pedigree and could not be housed and fed in the same manner as the hybrid communal area livestock. Potential beneficiaries had to construct a shed on their property to house the livestock from the scheme. They also had to collect cattle feed for it and feed it on the property and/or graze it away from the communal herd. This pre-condition worked as a constraint as some of the poorest people in the community could not benefit from the initiative. When the heifer calved, members had to give the calf to the project if it was a female calf to enable new members to benefit. If it was male, the member kept it to increase his/her herd. The project also established a central dairy milk collection point in the district and provided a donkey-drawn cart and vehicle. The cart went around the members' homesteads collecting milk for sale to the urban areas. The project also extended seed and fertiliser loans to members. The project started in 1984 with 15 members and grew to more than 100 in 1996. At its peak, the project was delivering 500 litres of milk daily to the National Dairy Board. This had since declined to 40 litres a day in 2005. The donkey cart service stopped when the donkeys became too old. The members could not mobilise the means to replace them. As a result, individuals were responsible for delivering their milk to the collection point. The project suffered from the economic crisis and the sanctions that the international community imposed on Zimbabwe. The government-subsidised veterinary services were suspended and commercial farming activities in the area were disrupted. There were four members of the dairy project in Ndamba – three single women and the wife of the *sabhuku*. Only one member still had livestock from the project in 2005. The rest had lost their heifers to livestock disease. There was one member of the heifer international project in Makuku.

Makuku village had larger and more diverse population than Ndamba. This limited the basis for co-operation. There were some sub-units of support in Makuku. One consisted of the remaining agnatic group sharing in cattle herding activities, visiting and supporting the ill, and transferring land between themselves. The immigrants assisted each other through their own networks they helped new immigrants to gain access to village membership and land. There was a burial society whose members assisted each other in times of bereavement. Six homesteads in Makuku belonged to this organisation. There was a group of beer brewers. They shared materials and

[44] NORAD sourced the heifers from the commercial farming areas.

cooking utensils and took turns to cook and sell beer so as not to flood the market. Three homesteads belonged to this group.

The *zunde* (see note 3) project for Mwanza was supposed to operate at the village level. The government tried to revive the concept in communal areas in order to enable communities to care for orphans and vulnerable people. The *muchinda* (sub-chief) was in charge of it. Despite the fact that land was available for the project in the ward, and that the government provided seed and fertiliser, the project had failed to take off. The *muchinda* failed to mobilise labour from the ward inhabitants. The State provided drought relief twice during the fieldwork period. The Ndamba community all shared equally in their disbursement. The Makuku community was so large that homesteads took turns to benefit from the drought relief. This caused conflict between the villagers as shown in Box 4.1

Churches provide some links for support (Hartnack, 2005). The people from both villages attended a variety of churches. The church assisted its members with fellowship, moral support and material support. The churches provided financial assistance, school fees assistance and food for children who attended the crèche. The Apostolic Church purported to heal their members through prayer.

The two villages derived domestic energy from firewood, solar power and electricity. The use of paraffin for lighting had ceased during the fieldwork period because of the erratic fuel supplies. Three homesteads had access to electricity. They were all male-headed homesteads in Makuku village. One single woman hearth-hold in Ndamba used solar power. The rest of the homesteads in the two villages relied on wood fuel. Ndamba villagers collected their firewood in their community forest and they had a thriving Fambidzanai NGO-initiated tree planting project, shown as the gum tree (eucalyptus) plantation on Map 4.2. The community provided land and the donor taught the women how to graft saplings. The community nurtured and harvested the vegetation for firewood and cattle feed. Makuku villagers bought firewood from the market or stole from the adjacent commercial farming areas. The people from Makuku walked long distances to collect firewood from the commercial farms. These people also risked arrest as they harvested the firewood without the commercial farmers' consent.

There were a variety of livestock herding practices in the villages. Ndamba village had fenced off some of its fallow arable land for exclusive use as a cattle paddock by Ndamba villagers. This is shown on Map 4.2. The land marked on the map as open pasture lying between the river and the homesteads is unused arable land. Ndamba villagers also used their large homestead allotments for grazing. In Ndamba, people with livestock pooled their cattle for collective herding. The influx of immigrants into Makuku was accompanied by the establishment of Makuku 2 and Makuku 3 on the community grazing lands. Makuku residents all complained about the lack of grazing which they claimed to be declining as the result of the influx of immigrants. Makuku residents also illegally grazed their livestock on commercial farms. Cattle straying onto commercial farms were impounded and the owners fined. Collective cattle herding arrangements excluded immigrants in Makuku. Some people hired others to herd cattle for them, whereas others preferred to herd their cattle themselves. The breakdown of cattle herding co-operation and the scarcity of grazing land in Makuku increased the costs for the people who owned livestock as they had to go further to graze their livestock and enter into arrangements that increased demands on their labour.

The political arena offered opportunities and threats for the people residing in the villages. The ZANU-PF party organisation was used for channelling government benefits. These included employment on food-for-work schemes. Youths were employed through the party as local police service officers. They derived allowances in return for selling membership cards and provided security during state functions. The politicisation of resources in the communal areas meant that political rather than needs-based criteria were used in the targeting of state transfers. This increased the vulnerability of the most desperate individuals. The people who participated in opposition party politics were marginalised from the state support and their homestead members were also ostracised.

4.8 Livelihood dynamics

The hyperinflationary environment had a negative impact on people's livelihoods. People lost assets, networks and capital on which they previously relied to construct their livelihoods. The villagers reacted by taking certain decisions to ensure their livelihoods. The next section describes the impact of this process on people's livelihoods.

There were fourteen men in Ndamba village during the fieldwork, ranging in age from 16 to 70 years. Ten of these men were below 30 years of age. The men indicated in interviews that they were in the village because they had failed to secure employment in the urban areas, growth point or the mines where they had wished to earn wages which they could use to develop their homesteads. The people in Ndamba reminisced about a by-gone era when resources were abundant. The men also reminisced how easy it used to be to secure employment with little or no education. The men's presence in the village increased the availability of agricultural labour. However, this labour could only be effectively exploited by families who had access to the other agricultural inputs.

The downward spiral in livelihoods started with the war which was followed by the economic boom of the 1980s soon followed by the implementation of the structural adjustment programmes, which coincided with the drought of 1991-1992. People remembered going for days without food, losing their livestock because there was no fodder and withdrawing their children from school. The younger men had a different life experience. Although they were more educated than their elderly parents, it had been more difficult for them to secure employment. Two of the men were employed in Harare but were retrenched during structural adjustment. They returned to Ndamba where they herded cattle and engaged in subsistence farming. Six of the young men had been working in the informal sector as carpenters, miners or builders. They all lost their employment during operation Restore Order and returned to the village. The rest had never been employed since they left school. The elderly men in Ndamba, who included the *sabhuku,* survived on subsistence agriculture, receipt of a state pension (which had been eroded by inflation), and sale of firewood, craft, carpentry and building.

There were 65 men in Makuku during the inventory. Four were pensioners who used to be in the civil service. Eight men were not working because they were unwell. Ten men aged between 33 and 66 had been retrenched from their work on the former commercial farms or from Harare during the structural adjustment programmes. There were 37 men who had never been employed. More than two thirds of those 37 men were aged below 25 years. Six men were employed during the fieldwork period as the school caretaker, two teachers, a tailor, a grinding mill attendant and a welder in the ward. The grinding mill attendant and the welder lost their employment when the informal businesses they were working in were closed during operation Restore Order. The few men who had migrated from former commercial farms continued to work on commercial farms. The rest combined subsistence agriculture with building-brick moulding, painting, gold panning and trading. The failure by men to control economic resources undermined their ability to enter into marriage, provide communal area based family members with remittances or secure funding to establish and maintain a rural homestead. This not only increased the number of single women and men in the villages but it also explained the high incidence of undeveloped homesteads and ill maintained infrastructure on the homesteads. The situation in which adult men remained in the villages with limited access to resources also increased the burden of care borne by single women. This will be illustrated in Chapter 7.

The women recalled the time when the soils were rich, they received adequate rainfall and they had enough money to mobilise agricultural inputs for the planting season. Back then, all homesteads had to construct a granary to store their harvest. Some people referred to the ease with which the former commercial farmers provided mechanised draught for free. The former commercial farms had also been a source of food. The people in the research area bought subsidised food from he farms during drought years. The women also reminisced about the good roads that once served the communal area. The roads were served by subsidised transport. The women recalled how they used to rely on those buses to transport their fresh produce to the market in Harare. The people also talked about how they financed the construction of their homes, their children's education, the building of their livestock herd and marriage with the money they derived from their agricultural activities. The villagers also reflected on how the liberation struggle destroyed infrastructure and disrupted people's investment plans. The men related how their wages had been diverted to fund guerrillas at the expense of developing their homesteads. The women related how the war had disrupted their schooling through the destruction of schools, roads, bridges and health facilities.

4.9 Conclusion

Mwanza communal area during the study period was characterised by unemployed people, poor infrastructure, low and declining agricultural output and limited opportunities for diversification. The impact of the shocks on the different households and people was mediated by their respective positions. Previously, people coped by liquidating stored assets, resorting to the market, receiving supplies from the former commercial farms and/or transfers from the State and non-governmental organisations. The situation in the study villages shows that all these traditional coping mechanisms were no longer available. The timing of the shocks (structural adjustment, drought, economic collapse, fast track land reform, operation Restore Order, HIV/AIDS and inflation) has been rapid, with the effect that people have not had time between successive shocks to rebuild their livelihoods. Agriculture, previously a relatively important source of cash through crop sales was no longer the main activity for all communal area residents. The decline in the contribution of agriculture to livelihood means that ownership of livestock, arable land and farming equipment are only relevant indicators of livelihood activities for some but not all of the people in the villages.

The vulnerable people were drawn from the different population groups. Children were vulnerable as they were losing parents and educational opportunities to prepare them for the future. The young men lacked access to opportunities to diversify livelihoods, establish homesteads and accumulate assets. This increased their vulnerability to participation in illegal activities such as gold panning and poaching from the former commercial farms. The fast track land reform resulted in the ejection of a large population of vulnerable people into the communal areas. These former commercial farm workers swelled the numbers of vulnerable people in the communal areas and supported the emergence of a land market.

The high demand for land and pressure from migrants put a strain on the traditional land access mechanisms. As a result, new and innovative modes, such as the market, have become increasingly important. However, the new access modes introduced new vulnerabilities for the old and new communities. The extent to which the change in land access modes influenced single women's land access and livelihood decision making was mediated by the specific governance contexts. Ndamba village governance reduced people's vulnerability by suppressing land demand. This has resulted in the continued availability of natural capital for residents and social security. The village increased people's access to donor-initiated resources because of its co-operative nature. They have benefited from seedlings, livestock, herbs, and the gardening project. This has increased people's diversification opportunities.

Makuku village's relatively more flexible and decentralised governance system resulted in a diversification of resource access modes which opened more opportunities for vulnerable people to gain village membership. The land market was one of the emergent ways. This heightened competition for resources and led to conflict and breakdown of the community-based social security. This in turn increased people's vulnerability as they had to obtain these resources on the market or go without. Although the governance context did not have an impact on the prevalence of morbidity and mortality, the specific local conditions determined the extent to which social security mediated individuals' vulnerability. Ndamba village was shown to be better organised to assist individuals who had suffered misfortune, whereas the Makuku community was increasingly losing its cohesiveness because of the high population turnover and migrant influx introduced by the informal land market. The various activities of the households show the various constraints that individuals living in communal areas face. The vulnerabilities vary by age, homestead composition, access to cash and identity in the communal areas. The livelihood portfolios show the opportunistic decision making associated with the rapidly changing and uncertain environment. The proximity of commercial farming areas and the growth point (both areas of high HIV/AIDS prevalence) make the population vulnerable to the disease. The high mortality and morbidity rates in the survey population confirm this. The high proportion of children being fostered on the homesteads also indicates an increase in the vulnerable population.

As the capacity to secure rural livelihoods through urban wage work declined, so did the communal area's capacity to subsidise urban livelihoods and provide long term security. Some of the men reacted to this by disposing of some or all of their communal area land. These decisions increase the vulnerability of not only the men but also of their wives and or children in the future. This is because of the loss of the rural land which has historically formed an important fall-back position. The men in Ndamba could not dispose of their land as the *sabhuku* did not allow that. This ensured at least that the land remained and provided security for the men and their families.

The high number of single men, women and different household formations shows that marriage is no longer the only way of organising economic and social relationships between men and women in the communal areas. This is also confirmed by the different ways in which the different people gained access to productive resources.

People gained access to productive resources through the market, relations and politics, that is, through institutions that are not necessarily traditional and/or patriarchal. This increased the opportunities for single women to access resources in communal areas. As we will see in the next chapters, the extent to which resource access deviates from the norm is still dependent on the local context. In this case, the *sabhuku* was one of the more influential institutions shaping land access. The emergence of new land access modes increased the diversification in the way that people access resources. The impact was not felt equally in the diverse populations.

The sustained pressure on livelihoods in the communal areas resulting from the collapse of urban wage sector has drastically altered the rural-urban connections for both men and women. The experiences of the villages show the changes also negatively affected the utility of land as an agricultural asset and a source of security has declined. This has been reflected in people's failure to develop the land and willingness to dispose of this asset on the market. The consequences of this for the single women and the general population will be examined in subsequent chapters.

Chapter 5

Leaving, remaining and living: single women's access to land and community[45]

5.1 Introduction

This chapter discusses single women's access to residential land in Mwanza communal area. Access to residential land means access to a community as well as a place to live. It is sometimes forgotten that access to land in rural areas includes both a place to live and a place to work, and that these are not necessarily the same. Socio-economic developments, changing rural-urban connections and the high incidence of deaths from HIV/AIDS is putting pressure on kinship relations and land availability, which in turn threatens single women's land claims. Single women with young children are more vulnerable. The chapter illustrates how single women continuously strive and strategise to respond to their constraints, and the limits of their responses in the face of severe structural constraints. The focus is on the decisions which the single women make in their different circumstances and the consequences for their residence access in communal areas. The chapter highlights the importance for vulnerable women of access not just to land as a resource but also to membership of the community in which the land is located. Finally, by exploring the strategies and paradoxes, the chapter shows women not as passive recipients but as active decision makers influence the outcomes following changes in their marital status.

This chapter is based on the 22 focus hearth-holds' residential histories over their life courses. The focus group discussions with village heads, Zimbabwe Widows and Orphans Trust (ZWOT) in Harare and commercial sex workers at the growth point also inform this chapter. The 22 hearth-holds and the 18 homesteads that they live on which were sampled from the two villages as explained in Chapter 2 are the focus of this chapter. Homesteads are the physical spaces within which hearth-holds are located in the village. As mentioned in Chapter 2, this study focused on the life histories of 22 hearth-holds that dwelled on 18 homesteads. The hearth-holds were either heads of homesteads or dependent on male-headed or female-headed homesteads. In Makuku, eight homesteads were hosting more than one hearth-hold. The 22 hearth-holds in the 18 homesteads studied are the specific subjects of enquiry. The concept of access is used to analyse how women gain and maintain residence as it acknowledges the importance of place, socio-economic factors, employment, vernacular market, relationships and identity in determining access to land, community and residence.

---

[45] Gender and Land Relations Project by African Institute for Agrarian Studies (under review).

5.2 The limits of marriage and custom

Hearth-holds gain residence on a homestead, and thereby community membership, through marriage, birth or the market. There are two types of officially valid marriages in Zimbabwe: a civil, formal or church wedding under the Marriage Act (Chapter 37) and a potentially polygamous customary law marriage under the African Marriage Act. The name of the latter act was changed in 1995-1996 to the Customary Marriage Act, Chapter 5:07 (Dry *et al.*, 1992; Vijfhuizen, 2002). All marriages in Zimbabwe are potentially polygamous (May, 1983). Customary marriages are not registered but acquire legitimacy through bride wealth payments and the presence of witnesses who were negotiators during these payments (WLSA, 1990; Vijfhuizen, 2002). The WLSA research on *Lobola*[46] in Southern Africa in 2002 concluded that, in all of the marriage types, the husband is considered the head of the family, women do not enjoy the same rights as their men and that marriage under all regimes undermines the socio-economic position of women (WLSA, 2002).

Customary marriage has tended to be the more common one among Africans in rural areas. Box 5.1 shows that there are at least nine different varieties of customary marriages in Zimbabwe. The customary marriage can be concluded and dissolved without any court formalities and is not necessarily recorded on paper (May, 1983; Vijfhuizen, 2002). This makes it easier for couples to terminate customary marriages. The 22 women that were studied respectively entered into diverse marriages which are highlighted in the individual case studies in Chapters 5 and 7.

Ideally, all marital forms aim to culminate in the full exchange of bride wealth and formalisation of marital relations. This is supposed to give a woman status, shelter and security, and ensures the legitimacy of her children and their rights to their father's lineage benefits (Mvududu and McFadden, 2001:205). *The kutizira, matorwa, kuzvarira, kusenga bere, chigara mapfihwa and kugara nhaka* (1, 2, 3, 5, 6, 7) are virilocal marriages, i.e. the couple live in the man's village of origin. Marriage types 4 and 9 do not necessarily result in a virilocal marriage. Although marriage forms 3 and 8 are illegal, they are still practised and were encountered in the study villages. Again although illegal, the land market is another way through which people access homesteads in Zimbabwean communal areas. The people who buy land cannot strictly be described as being in virilocal marriages. Here they are referred to as 'market' to distinguish them from the marital residence gained through kinship in communal areas. The diversity of marital arrangements results in different decision making and outcomes for the various women who enter into each respective union. This means that it is not possible to generalise about the relation between customary marriage and land access in the communal areas of Zimbabwe. It is also not possible

[46] *Roora/Lobola*: the process where the family of the man makes payment to the family of the woman in the process of marriage.

Box 5.1. Types of customary marriage in Zimbabwe.

1. *Kutizira*: elopement where a girl moves to the homestead of her future husband without the consent of the approved channels. This should be followed by the husband's family initiating marriage negotiations with the bride's family to formalise the relationship (Kuper, 1954:21; Udry *et al.*, 1992; Vijfhuizen, 2002).
2. *Matorwa*: the girl is collected from her parents' homestead by the husband's aunt (Vijfhuizen, 2002).
3. *Kuzvarira/kuroodza/kwendisa*: poor families who have difficulties betroth a young daughter to a family that will provide bride wealth, which enables the family to survive. This is prohibited by law (May 1983:45, Mvududu and McFadden, 2001:208; Vijfhuizen, 2002).
4. *Kutemaugariri*: when a poor man who is not able to pay bride wealth will live and work at the wife's homestead providing his labour in lieu of bride wealth (Kuper, 1954:21; May, 1983; Vijfhuizen, 2002).
5. *Musengabere*: the practice whereby a man rapes a girl in order to get her to marry him. This is a forced marriage and is forbidden by the government (Vijfhuizen, 2002).
6. *Chigara mapfihwa*: the practice whereby a deceased woman's sister moves in with the widower as a replacement for the wife (Vijfhuizen, 2002).
7. *Kugarwa nhaka/Levirate*: a woman is looked after by the brother of the deceased husband (Mvududu and McFadden, 2001:206; Vijfhuizen, 2002). WLSA research on inheritance in Southern Africa concludes that it is misleading to refer to a widow's remarriage within the deceased husband's family as widow inheritance, as women have a choice of whether to accept the union or not (Aphane *et al.*, 1995:42).
8. *Ngozi* type marriage: *Ngozi* is believed to be the avenging spirit of a person who was killed. People believe that such a spirit causes illness and other misfortune. The family suspected of a killing has to give a daughter to the family whose member was killed. That woman will be married to a male relative of the deceased to bear children. People believe that, through such a 'ghost marriage', the wish of the angry spirit is fulfilled and that this will prevent illness and other misfortunes. These ghost marriages are forbidden by the government but they still occur (Vijfhuizen, 2002:23).
9. There is another alternative marital form known as *mapoto*: a type of a co-habitation, this union is not based on male dominance but on negotiated relations between a man and a woman. Since there is no bride wealth, *mapoto* wives do not have obligations of women in patriarchal marriages, such as moving to the man's village. These unions are easier to dissolve than formalised relationships (Mvududu and McFadden, 2001; Muzvidziwa, 2002).

to generalise about consequences for the women who enter into different marriages. The diversity of the consequences will be highlighted in the specific case studies where relevant. While marriage is an important relationship in determining women's access to land in communal areas, the chapter also illustrates that it is neither the sole nor most important determinant of women's village residence and community membership.

5.2.1 Alice: losing a marital home

Alice was a single woman hearth-hold who was living with her *mapoto* husband in Ndamba when the fieldwork started in 2004. By the end of the fieldwork, Alice had relocated to her deceased parents' village which was adjacent to Ndamba the village. In order to reconstruct Alice's residential history, it is important to interrogate the whole history of the homestead that she occupied for three years in Ndamba village.

The Ndamba *sabhuku* occupied a piece of land for his homestead, which he later abandoned because of the poor soil quality. The *sabhuku* later allocated the infertile land to a 36-year-old man who had worked for him. The 36-year-old married and built a kitchen for his wife, sank a well and established an orchard. The marriage produced two daughters. The man's wife became ill and moved to her parents' village for care. She died in 2003 from tuberculosis. The two daughters went to live with their maternal grandmother. The man subsequently entered into a *mapoto* union with Alice, a 33-year-old woman who was an unmarried mother of two. Alice's children were aged 11 and 8 years, respectively, when fieldwork was conducted in 2005. Prior to the marriage, Alice lived in her natal village, Mapfumo, adjacent to Ndamba. The *mapoto* marriage between Alice and the man from Ndamba produced a daughter. Alice moved onto the husband's homestead in Ndamba and left her other children at her natal village. Alice's husband died from meningitis when their daughter was two years old. The Ndamba *sabhuku* instructed Alice to find the man's relatives so that they could repatriate the man's body for burial in his own village of origin. The man's parents then came and buried their son in Ndamba village. They could not afford to repatriate his remains for burial in their own village. The man's parents did not acknowledge Alice as their son's wife because she was not formally married. The man's father took the radio and bicycle, locked the house and took the keys with him. Alice's relatives took her to her natal village so they could comfort her. They also felt that she was too young to live on her own. Alice, now a widow, tried to maintain the marital residence by watering the vegetables and cleaning the place.

The Ndamba villagers started using the homestead as a common property, by grazing their livestock and helping themselves to the fruit from the orchard. It was difficult for Alice to secure this garden as she no longer resided in Ndamba. The deceased man's father later returned and sold the fencing, roofing and all building materials which he could strip from the homestead. The land was returned to the Ndamba *sabhuku*. The *sabhuku* subsequently allocated this homestead to his son who lived in Harare. Alice relocated permanently to her natal home.

In an interview, Alice described how from the start she had been excluded from the village. She had been marginalised from the women's garden co-operative and other activities in the village. During her husband's illness, the family never received any assistance from the village health worker. Their name had not been included on the list of food-aid beneficiaries. A month after the death of her husband, somebody else died in the Ndamba village. Alice went to pay her condolence subscription and was told that her name had been struck from the village register as she was no longer considered a member of the community.

In her natal village, Alice resumed occupation of her late parent's homestead, which she had ceded to a younger sibling when she moved to the marital homestead. The sibling was 15 years old when Alice returned to her parents' homestead. Alice lived there with her three children, an orphaned niece and her own brother and younger sister. The brother, a school dropout, was an active opposition party supporter. As her parents' village was adjacent to Ndamba village, Alice invested in cultivation in her natal village, received food aid and government relief through membership of that community and was a senior member of the garden co-operative in her natal village. She had maintained this position for the duration of her marriage. Her harvest failed as she had spent most of the work time looking after her ill husband. She survived by doing piecework for fellow villagers in exchange for food.

Alice's eight-year-old son fell ill during the fieldwork period. She took him to her faith-healing-based church for prayers, but he did not improve. She borrowed money and took him to the clinic where they were referred to the hospital. Alice did not succeed in borrowing money from the village to take her son to hospital as she had a bad credit record. A neighbour asked her to provide some 50 bundles of thatching grass in exchange for the money to pay her son's hospital expenses. Alice worked for a week and produced seven bundles. Her son got worse and died before she had raised enough money to take him to the hospital. When the son died, her church, which normally assisted her with food and money, refused to participate in the funeral because she had taken the child to the clinic against their advice. Alice also spent time nursing an ill neighbour who was her major source of employment.

Alice's experiences show that access to a homestead and community membership is important for communal area livelihoods. It is from the homestead that people can negotiate membership of the community. However, though she successfully negotiated access to a homestead in Ndamba, Alice was not accepted as a member of the community into which she married, despite the fact that she had taken up residence in that village, because her husband was an immigrant, and also the union between Alice and her husband was not a socially sanctioned one. Consequently, the traditional support networks did not extend to people in *mapoto* unions. As a result, Ndamba village restricted her access to community level support. Alice's experience illustrated that it is within the wider community that hearth-holds have to negotiate access to livelihood capitals for managing risk. These include labour, physical, moral, financial support and, if necessary, land to farm. The government and non-governmental organisations channel seed, drought relief and employment opportunities through the community structures. Other residents of the community, (adult children, the church, the State or the extended family) potentially provide support.

The social support systems were breaking down because of the economic crisis and mobility induced by the changing rural-urban connections. The state health system was failing to deliver, so people had to mobilise their own networks and financial resources to deal with illnesses. Alice's example shows that, in order to benefit from the community, hearth-holds have to be recognised as members by the other people in the community. There may also be obligations and conditions that community members have to observe in order to be eligible to benefit. Although agriculture is no longer the main livelihood activity, residence in the communal farming area is important for people as it is from there that they negotiate access to non-agricultural income earning opportunities in the communal areas, urban areas, commercial farms, mines and the growth point.

5.3 The formation of single women hearth-holds

The single women hearth-holds' life histories revealed that these women have quite diverse life-histories and living conditions. Single women hearth-holds are composed of divorced, widowed or abandoned women. These categories are not mutually exclusive as the case studies show, an individual hearth-hold like Alice cited above can have at different points in their lives become widowed, abandoned and divorced. Married women whose husbands die before them become widows. Sixteen of the single women in the in-depth sample were widows at the time of fieldwork. When spouses separate after bride wealth has been paid, it is called a divorce. There were five divorcees in the in-depth sample. Some of the women had had children with men but had not married them.

Sixteen of the twenty-two hearth-hold heads in the in-depth sample had been widows at some point in their lives. The total number of women who had ever been divorced was eight. Five women had had illegitimate children. There were women who in their lifetime had become widowed or divorced more than once or widowed more than once. Two women in the sample, both in their thirties, had four and six children, respectively, fathered by three different men. Both expressed hopes of getting married in the future. The life histories of single women also demonstrate the fast-changing circumstances in which they lived. The marital status of three single women-hearth-holds in the sample of 22 some changed during the fieldwork period. One was Alice who was a single mother, a married woman and a widow over the time span of three years. In another instance, a single woman of 19 was married in a *kuzvarira* marriage and subsequently divorced. There was a serial divorcee who was in a *mapoto* marriage that ended during the course of the fieldwork. The last example was a formerly abandoned woman who when the research started was in a *mapoto* marriage to a man who later succumbed tuberculosis leaving her a widow. Thus hearth-holds were constantly negotiating homestead access because of changing conditions.

A married woman normally gains access to a village and place of residence through marriage, i.e. she obtains access to a homestead following her marriage and ties to her husband. The study will illustrate that this is not always the case. In the present study, this type of formerly married woman was to be found in both villages and constituted the majority of single women in Makuku. Former male commercial farm workers move into or retire into communal farming areas, and, although not born in the communal areas, these men marry women from the host community and successfully negotiate for land on which to establish their homesteads. This is akin to a *kutemaugariri* type marriage as the husband moves to the wife's village on the marriage. These marital homesteads are located in the woman's natal village, i.e. among the woman's kin. Such homesteads were encountered in both Ndamba and Makuku during the study period. Chapter 4 highlighted that some people purchased land in Makuku in order to establish a communal area home closer to Harare and surrounding commercial farms. This was a response to the high commuting costs of maintaining rural-urban connections. While they are employed and residing in the cities and on commercial farms where they work, these purchasers' homesteads remained unoccupied in Makuku. Some single women in Makuku were occupying homesteads as caretakers for absent homestead owners. These will be discussed in the next chapter. The sample had some single women who entered the village through the purchase of homesteads on the market. There was one single woman who purchased a homestead in Makuku in her own capacity. However, women do not always have this possibility. A study in Svosve communal farming area concluded that the *sabhuku* discriminated against single women when selling land (Chimhowu and Woodhouse, 2006b).The other single women hearth-holds resided in homesteads purchased by their deceased husband and/or adult children, respectively. The women at the growth point rented houses from the market. The diverse ways though which married couples access homesteads in communal areas has given rise to more divergent rules on the governance of property on marital dissolution. Examples include but are not limited to the allocation by the *sabhuku* to a married man from the village, allocation by the *sabhuku* to an immigrant man who married into their village, *mapoto* and/or purchase on the informal land market. The experience of single women in various marital statuses will be used to highlight the main factors that influence single women's decision making in residential space.

Although marriage is historically one of the commonest ways of gaining residence on a homestead, this can be threatened by the death of the spouse, divorce and the failure of the man to complete the marriage formalities. Traditionally, a woman widowed when she is past childbearing age comes under the care of the head of the family whose duty it is to see that she has a means of survival, i.e. land to cultivate. A widow's continued access to productive assets in her late husband's homestead sometimes is contingent upon her remarrying a brother of the deceased husband (WLSA, 1994, 1997) in a *kugarwa nhaka* type marriage. In this study, the focus group discussions concluded that, in most virilocal marriages, family courts were convened to determine the future of a widow with regard to her late husband's homestead and arable land. The family courts usually consisted of the deceased husband's family members and the widow herself. The *sabhuku* might attend the family court but in the capacity of observer rather than as an active participant. The outcome for widows was a function of residency type, marital status, life course stage/age, economic position/class, the woman's agency, the woman's health status and children. The specific experiences varied with individual women's circumstances and at times contradicted the findings of the focus group discussions.

Folta and Deck's (1987) study in a communal area in Zimbabwe revealed multiple patterns of inheritance of property utilised in the case of widows. Either widows inherited the property of the husband or it was shared among the wives and children, or among the husband's family (1987:330). Cases of *masabhuku* dispossessing widows have also been noted in other studies in the communal areas of Zimbabwe (Izumi, 2006:30). Focus group discussions in the study villages concluded that, in a virilocal marriage when the male spouse died, the couple's possessions were distributed in an inheritance process. The inheritance process involved the decision making that the family entered into concerning the deceased man's property and family. The process ranged from a formal discussion involving the deceased's family, the wife and/or her relatives to the situation described in Alice's case where the widow was marginalised in practice by both the *sabhuku* and her deceased husband's family.

The property under consideration usually included urban immovable property, radios, bicycles, ploughs and livestock. The governance of urban immovable property falls under the jurisdiction of formal law and is administered in urban centres.[47] Four widows in the two study villages had retained access to urban immovable property. None of them had completed the legal formalities to have the property transferred to their names. The contraction of urban wage earning opportunities reduced communal area residents' access to remittances and also undermined opportunities for men to earn an income. As a result, the rent from the urban immovable property enabled

[47] The process of inheriting urban immovable property is a cumbersome one for widows because of the historical discrimination and lack of clarity resulting from legal pluralism and limited access to institutions that administer state law. This renders the property vulnerable to repossession by the local authority or children and relatives of the deceased man who may be in a better position to navigate the practical and financial obstacles than the widow (Paradza, 2007).

single women hearth-holds that had control of such property to maintain rural-urban connections which underpin communal area livelihoods.

The focus groups reported that normally the deceased man's relatives were entitled to his clothing and property that belonged to the extended family. The property usually included agricultural equipment and livestock that the deceased man had inherited but did not include land and/or housing. Some of the widows in the sample retained the radios, wheelbarrows and bicycles. However, most of these were broken down and no longer in use. Alice's experience provides evidence of property grabbing[48] by the deceased husband's relatives. This was the only case of property grabbing encountered in the study area. If a widow is allowed to continue using the assets such as the homestead, land, agricultural equipment and livestock, she is more likely to remain in the homestead than one who is stripped of the physical assets. The focus group discussions concluded that women who had established an independent homestead with their husband and developed it were more likely to remain in the marital village on widowhood than those who had not yet developed their homesteads. Independent homesteads were normally allocated to domestic units after a period of time, depending on availability of space and the child-bearing status of the woman. This disadvantaged younger widows as couples were struggling to gain access to independent homesteads and/or establish their own homesteads on land allocated to them. However, Alice's experience, where she had an independent homestead and had borne a child but was unable to retain access to the homestead following the death of her husband, contradicts this assertion.

The focus group discussions also revealed that being accepted as a member of the community was an important factor in hearth-hold's decision-making. The community provided support, assistance with child care, moral support, assistance with agricultural tasks and recognition of the hearth-hold as a member. Access to money reduces a widow's need for assistance. This is because the widow can use the money to hire people to perform tasks that would have otherwise been performed by her now deceased husband. The tasks include ploughing, repair of the homestead and/or excavating a well. The focus group discussions showed that widows sometimes experienced hostility from wives of the deceased husband's brothers who felt threatened by the widow. As a result, these women withheld support and assistance from the widow. Alice's experience shows how a whole village collaborated to exclude her hearth-hold from the community by their failure to acknowledge her as a recipient of government aid, their treating her homestead as open pasture upon which they could allow their livestock to graze and their refusal to allow her to contribute to the village burial society. Hearth-holds that find themselves in such situations are more

[48] Property grabbing is the process whereby an individual is forcibly evicted from her home by other family members, traditional leaders or neighbours and is often unable to take her possessions with her (Izumi, 2007:12).

likely to leave than remain in a marital village. Some widows in the study remained in the marital villages with neither money nor social support because they lacked alternatives or wanted to protect the residential rights of children who had moved to urban areas. Specific examples are highlighted in the case-studies described later in the chapter. A woman's age influences the hearth-hold's material status and the status of adults in the community, their obligations and the help they may receive (Izzard, 1985). A woman widowed in her elderly years is more likely to remain in the marital village than a younger woman, because her experience and senior status increases her chances of successfully negotiating for continued community membership. All the younger widows with virilocal marriages encountered in the sample had left the marital village when their spouse died. Alice's experience shows that it is not possible to generalise the decision making of single women. Even though the focus group discussion indicated that elderly widows (postmenopausal) were more likely than younger ones to remain in the virilocal marital village, the case-studies show that elderly widows also relocated following the loss of their spouse.

A woman widowed after she has borne and reared children is more likely to decide to stay in the village than one who has either no children or very young children, because younger children need assistance and support which a young woman would have a difficult time mobilising. She would have to resort to her own kin who are not the agnatic kin of the children. Younger women who are still in the reproductive phase also leave marital villages to pursue other opportunities to remarry (Vijfhuizen, 2002:15, Izumi, 2006). Women who are still in the reproductive stages normally stay on in a marital village if they decide to marry one of their husband's brothers in a type 7 marriage (see Box 5.1), popularly known as *Kugarwa nhaka* or levirate. The *kugara nhaka* practice in Zimbabwe has been declining because of changing norms, the economic independence of women through receiving the husband's pension and increasing knowledge about the spread of HIV/AIDS (Vijfhuizen, 2002; Huisman, 2005). The next section focuses on specific women's circumstances to illustrate the ways in which single women gain negotiate and maintain community membership and residence in communal areas.

5.4 Remaining: women who remained in the marital residence after the end of their marriage

5.4.1 Mary: maintaining and consolidating a prosperous homestead

Mary was a widow who lived on a homestead that Mary and her husband were allocated by the *sabhuku* upon their marriage. She was living in Ndamba village in 2005. She had already been a widow for 12 years at that time. Mary had the best developed and maintained homestead in Ndamba. There were five brick and asbestos buildings, two toilets, a well, a vegetable garden and a large granary. She lived at the house with her two adult sons and a grandson. One of the sons was married and his wife lived there with Mary. This married son and his wife had been allocated a place to develop a homestead by the *sabhuku* but they lacked the labour, finance and building materials to do so. Mary's younger son was still single and therefore not entitled to a homestead in the village. Mary and her husband were allocated the homestead by *sabhuku* Ndamba in 1977. The husband worked in the urban areas while Mary looked after the homestead and farmed. The husband constructed two buildings. Mary's husband died in 1993, and she continued to live on the homestead after her husband's death. This was after she chose not to enter into a *kugarwa nhaka* marriage with one of her deceased husband's remaining brothers. Mary was the head of this homestead and in charge of decision making and control of the assets. Mary inherited the plough and radio that had belonged to her husband. The husband's family shared the clothing. Mary started going to South Africa to buy clothing for resale in Zimbabwe in 1997. This was a highly profitable venture and was her main source of income. Using the proceeds of the business, Mary added another building to the homestead and invested in solar power. She also completed the construction of her house in Harare, which she and her husband had started while he was alive. She leased that house in exchange for rent. Mary's unemployed son and the daughter-in-law helped her to keep the house and look after it when she was absent. Although Mary was in charge of the homestead, the *sabhuku* retained control over this and other single women hearth-hold headed homesteads. This is explored in Chapter 6. Mary's experience shows how a communal area homestead was developed and maintained using resources that were secured through rural-urban connections. The case also illustrates how the connections changed over Mary's life course and the way in which she repositioned herself to maintain access to financial resources. Initially the homestead relied on the wages from Mary's husband to develop the homestead and invest in agricultural activities. Thereafter Mary herself exploited this connection by trading goods and investing the proceeds into the development of the homestead. This enabled the hearth-hold to develop and maintain her homestead and livelihood in the communal areas after the death of her husband.

5.4.2 Lizzy: maintaining a marital homestead with no income

Lizzy was a widow who lived on a homestead that had been allocated to her and her husband by the *sabhuku* when they got married. Lizzy was at least 65 years old in 2005. She lived in Makuku village. Her deceased husband was an original inhabitant of Makuku. She never went to school. When she married, Lizzy and her husband were allocated a homestead by the Makuku *sabhuku*. The couple lived and worked in Harare for more than 30 years. They constructed a house in Makuku. The couple also acquired a house in Harare when the government converted the African rental housing to ownership for sitting tenants after independence in 1980. The couple had a son. When the couple retired to Makuku, their grown-up son married and took up residence in their urban house with his wife and children. The son was not employed and offered no financial support to his parents. Lizzy's husband drowned in 1999. Lizzy blamed her husband's brother and his family for his death. This soured relations between the widow and her husband's family. She remained in the marital village living alone on her homestead. Lizzy had a homestead that was not fenced. The homestead had a three-roomed brick and asbestos structure, a hut that no longer had a roof and a dried out well. There was no toilet.

Lizzy inherited her husband's bicycle, plough, livestock and homestead. She lost all her livestock to the drought. At her age, she had no money or grain to hire people to repair her fence; this rendered her crops vulnerable to livestock. She struggled to maintain the fence as she lacked the skills and materials. The holes in her roof were filled with bits of paper; loose masonry was secured with maize stalks and pieces of firewood. She stored her clothing in upturned metal containers to protect them from rats. She had a banana orchard on her homestead but had never benefited from it as the fruit was eaten by cattle. Her homestead well dried up and she could not afford to have it excavated. Lizzy relied on neighbours for water. The toilet that she and her husband constructed fell apart. She used bits of wire to secure her house in her absence. Almost all her windowpanes were broken and had been replaced with cardboard. When her son fell ill, she went to visit him in Harare. During her absence, her home was broken into and the plough and bicycle she inherited from her husband were stolen. Lizzy opted to stay on in the marital village to maintain her son's claims to the homestead. She survived on handouts from other villagers and the little that she managed to produce on her homestead plot. Lizzy's case illustrates the role of urban-rural connections in facilitating the communal areas residents' access to finance. The finance was used to develop the homestead to which the couple retired to after their working life. The homestead suffered when the rural-urban connections were severed when her husband died. The urban house a potential source of financial income was occupied by Lizzy's son. The labour (son) that Lizzy's homestead exported to urban areas failed to remit any money. As a result, Lizzie could not maintain her homestead infrastructure or invest in agricultural activities.

Lizzy and Mary's cases describe the experiences of women who have gained community membership and access to residence through marriage. Both women stayed in the marital village after the death of their spouse. The two women retained their husband's homesteads when they became widows. They both had established independent homesteads when they lost their husbands. They also had grown-up children, more importantly, sons, which strengthened the women's claims to the marital homesteads. However, the two women's pathways demonstrate different vulnerabilities. Although the communities accepted the two widows as members, the village offered limited assistance. The two women were vulnerable in different ways. Mary's strategies to reduce vulnerability included living with her son and his wife, maintaining a source of income by repositioning herself in order to maintain the rural-urban connections and successfully leasing the house in Harare. The cash income increased her autonomy and reduced her dependence on the husband's kin.

The presence of her son and his wife provided Mary's homestead with security in her absence. Lizzy's options were limited by her advanced age and the decision to allow the son to occupy the urban property. As a result, Lizzy's urban connections did not result in the flow of financial capital but were characterised by the rural area providing security and a fall back for the son who had migrated. Without the financial lifeline from either rent or remittances, Lizzy became more dependent on traditional assistance. However, the relations on which she would have drawn for support had become strained when her husband died, because of the accusations. The two women both had adult sons who had failed to secure employment. The living arrangements of the women and their sons rendered Lizzy more vulnerable. Mary's residential arrangements increased her access to her adult son's labour and household maintenance. Lizzy's vulnerability increased when she went to visit her ill son as she lost some of her possessions, whereas Mary's living arrangements ensured that her homestead was secure in her absence. Lizzy's experience shows how even when they retain the marital assets, widows remain one of the most vulnerable populations in communal areas. Lizzy's increasingly vulnerable position shows the importance of rural-urban financial flows for the maintenance of rural livelihoods.

5.4.3 Ginny: maintaining and re-negotiating homestead access

Ginny was a widow who continued to live on a homestead that had been allocated to her and her husband when they got married. Ginny was about 63 in 2005. She was one of the poorest hearth-holds in the sample. Although Ginny still controlled the arable land that had been allocated to her husband upon their marriage, she did not fully utilise the land as she lacked inputs, labour and was hosting an ill adult. She survived on handouts and collection of wild vegetables and occasional work for other people. Ginny had a son out of wedlock in her teenage years. She later married into Ndamba village where she and her husband were allocated land to develop a homestead. They developed the homestead using proceeds from Ginny's husband who was employed as a wage labourer in town. Ginny used the wages to hire labour to work their agricultural land. The marriage produced six other children. Ginny's husband died after all her children became adult. Like Mary, Ginny elected not to enter into a *kugarwa nhaka marriage* and she continued living in the marital village. Ginny could not access her deceased husband's pension as her marriage was not formally registered. During the fieldwork period, Ginny's eldest son lost his wife to pneumonia. He too became ill and lost his job. He moved in with his mother so she could take care of him. The Ndamba *sabhuku* resented this arrangement and initially encouraged the elderly widow to take her ill son to his father's village. Ginny ignored the *sabhuku*. As the son's condition deteriorated, the *sabhuku* offered Ginny bus fare to enable her to transport her son to his father's village. The *sabhuku* was worried about the social cost to the village of not only caring for the ill man but also the funeral expenses. The ill man eventually died and had to be buried in his mother's marital village. Ginny received minimal assistance from the village residents to the extent that there were no meals for mourners. People congregated on the homestead just for the burial rites and dispersed immediately after. The case shows the increased vulnerabilities of single women in a community whose social networks have been weakened by economic crisis and HIV/AIDS. The desperation of the man who had to follow his mother to her new marriage village highlights the vulnerability of hearth-holds' children. The son's presence undermined Ginny's claims to community assistance because the son was not a blood relation of the villagers. The *sabhuku* withheld assistance and tried to persuade Ginny's son to leave the village. Ginny successfully resisted attempts to stop her from looking after her ill child. Ginny's hearth-hold's financial rural-urban connections were severed when her husband passed away. Even though her children had left home, none of them sent her money to invest in her land and/or the maintenance of the homestead. This made Ginny entirely dependent on the handouts and the little she produced with her own labour from her arable field. However, even though the homestead lacked a constant stream of finances from the urban areas, it still served as a fall back for her son when he fell ill which confirms the enduring position of the communal area as providing security for urban migrants. The extent to which the rural homestead can provide insurance for the ill and aged is also dependent on the nature of financial flows between the two. Without a sustained link, it becomes difficult to provide care for the ill and/or to meet funeral expenses from the rural home.

5.5 Leaving: hearth-holds who left the marital homestead after the end of their marriage

With the exception of Alice who relocated from the study villages to an adjacent village, the experiences of women who left the study villages after the end of their marriages could not be captured from them. The study relied on a combination of the stories of those who remained in the study villages, such as their mother-in-law, focus group discussions and women who lived in the study villages who had returned from their marital villages to settle in the study villages. The sample of 22 had eight women who had returned to Chikwaka communal areas following the end of their marriages. These were interviewed to ascertain the processes and circumstances that lead a hearth-hold to leave their marital home. One of these was Alice. The hearth-holds' accounts confirmed that the widows left because they were ill, too young, childless and/or had no means of sustaining themselves. There were some women whose decision to leave the marital village was made by their parents. Some women left the husband's village immediately after the burial of their husbands. Normally, the women would be accompanied by their parents who claimed to be taking them to their birthplace for 'comfort during their bereavement.' Such women rarely returned to their marital villages. Parents of such women claimed to have perceived hostility from the daughter's late husband's family, poverty in the marital village or bride wealth outstanding. Some claimed that their daughter was *too young* at the time of her husband's death to continue living in the marital village. A study in three communal areas of Zimbabwe found that widows were commonly accused of bewitching husbands who had died of HIV/AIDS (Izumi, 2006). More often than not the end of the marriage also signalled a change in rural–urban connections for the women most of who relied on remitted wages of a husband to establish and maintain a homestead. The single women hearth-holds had to seek alternative means to earn an income and/or relocate. The next section uses experiences of single women hearth-holds that left their marital residence following the change in their marital status.

5.5.1 Tracy: hearth-hold continuously renegotiating residence

Tracy was a widow who had returned to live in Ndamba after her two marriages ended with the death of her respective spouses. Tracy was living in Ndamba village and was approximately 70 years old in 2005. In 2005 Tracy was living on a homestead that was originally allocated by the *sabhuku* to Tracy's deceased parents. She also had access to the arable land that went with this homestead. She never went to school. Tracy's main assets were her, health, pottery skill, the relationship to the *sabhuku*, membership of Ndamba village and host community, the control of the homestead and agricultural land and a few chickens. She had no livestock. The Ndamba *sabhuku* was her nephew. Around the 1950s, when she was a teenager, Tracy married a man from Murehwa, a communal farming area located to the North of Goromonzi. The couple lived in Harare and worked as domestic workers for ten years. Tracy's husband died before the couple had developed their rural homestead. Tracy entered into a *kugarwa nhaka* marriage and married her deceased husband's younger brother and they established a homestead in Murehwa communal area. Tracy had three children with her new husband. When her second husband died, Tracy was blamed for his death. She left the village in Murehwa and her children after being threatened by the husband's family. She went to live in her natal village in Ndamba with her mother. After some years, her mother died and Tracy remained on the homestead.

When Tracy's sons grew up, they bought a homestead in their father's village and invited Tracy to come and live with them. Tracy left Ndamba and went to live in Murehwa with her sons[49]. She lived there for seven years and had her own house and some land to farm. She did not get on with her daughters-in-law who also had their homes on the homestead. After seven years, Tracy had to return to Ndamba to look after her widowed sister who was ill. When the sister died, Tracy's children barred her from their homestead in Murehwa. They accused her of bewitching them. Tracy lived at her late sister's old house before moving back to her parents' homestead in Ndamba. In 2005, she was living on her deceased parents' homestead. There were

[49] Tracy's children wanted to look after their elderly mother instead of leaving her to depend on her natal family as this reflected badly on them. They bought one homestead but it was for the family's use. Tracy would have been living there with her daughters-in law while her sons worked in urban areas.

two mud and pole huts, which had been constructed by her now deceased father. Tracy was entitled to live on this homestead even though her nephew the *sabhuku* retained the overall control over it. The well and toilet had both collapsed. The fencing was poorly maintained rendering her garden vulnerable to livestock. She lived alone on the homestead in her natal village. She received no support from her children. She thought that this was because they accused her of witchcraft. Tracy made clay pots for sale. As she was illiterate, she relied on young people in Ndamba to assist her with the pricing of her pots. She tried to practice organic[50] maize farming but complained that it was labour intensive. She was also employed at the spirit medium's house to brew beer for rituals. Shona cultural beliefs prescribe that only post menopausal women are involved in the preparation of this beer. These women are provided with food during the time they were working at the spirit medium's. Tracy had an herb garden but the herbs died during her absence and also in the dry season, as she did not have a domestic water source. In May 2005 she went away for a funeral and was absent for two weeks. During that time, the *sabhuku*'s children looked after her property.

Tracy's case shows how death and illness can undermine a hearth-hold's residence and community membership. Tracy tried to secure her interests by marrying the deceased husband's brother. This is a common strategy (Folta and Deck, 1987:330). The witchcraft accusation that followed the death of Tracy's second husband cost her access to the marital residence and the support of all her children. Tracy's experiences illustrate the extent to which beliefs can destroy relationships, undermine social security and increase a hearth-hold's vulnerability. Witchcraft accusations, which are often levelled at women, undermine the advantages of rank that women gain with age. In her case this cost her support from her children. Tracy had to rely on her own networks in her natal village. Age, which is so often seen to disadvantage women, proved to be an asset to the elderly woman as it enabled her to gain access to income earning opportunities through the spirit medium in Ndamba. The case shows that even support from adult children is conditional upon a hearth-hold maintaining cordial relationships with the children. The case also shows that women's domestic responsibility to care for the ill may cause them to move and undermine their homestead access. Tracy returned to her village of birth where her community membership and homestead access were restored. In her natal village, Tracy was able to benefit from the community. This came in the form of co-operation, assistance with pricing of her pots, participation in NGO projects, employment and security for her homestead by the *sabhuku*'s children during her absence. This was in contrast to Lizzy whose home was ransacked in her absence. In her natal village Tracy coped with the limited support by trying to be self-sufficient and reducing her dependence. Rural-urban connections in Tracy's case ebbed and flowed as she was initially

[50] This is a practice where, instead of using commercial fertiliser, the farmer relies on leaves, cow dung, etc. that occurs naturally in the environment.

employed herself and lived in urban areas and changed to the time when she was a recipient of her children's remittances which she used to finance her agricultural activity. After the ties with her children were severed, Tracy found a way to access financial capital through the pottery business and working at the spirit medium's. Although she managed to meet her other needs, Tracy's activities were inadequate to invest in the maintenance of her homestead infrastructure.

Negotiating access to residence and Community after divorce

There is no general rule for dealing with divorced women in the communal areas of Zimbabwe. A study based on a single village in Manicaland province concluded that divorcees returned to their natal village, cleared their own sites for settlement within the husband's village or were allocated land by the local authorities (Vijfhuizen, 2002:44). Others concluded that most divorced women were forced to leave the homestead (Andersson, 1999; Huisman, 2005; Goebel, 2005). A study conducted in Zimbabwe[51] in 2000, found that 40% of the older single women, 35% of divorced women and 15% of widowed women lived with their parents (Hindin, 2002). Another study concluded that divorcees who left their marital homestead were often destitute. They lost their children, land, homestead and income (Goebel, 2005:128). Yet another study concluded that divorcees were usually granted a piece of land to work for themselves (and their children) but remained heavily dependent on social networks (Huisman, 2005:256).

5.5.2 Rita: destitute divorcee conditionally accepted into natal village

Rita was a divorced woman who had returned to Ndamba village after her marriage had ended in divorce. Rita lived alone on a homestead in Ndamba village. She had access to an arable field but owned no livestock. The cooking utensils that she used had been left there by Rita's deceased aunt who had occupied the homestead before her. Rita was born in 1957. Rita was 48 years old in 2005 and lived alone on a homestead in Ndamba. She was the *sabhuku*'s stepsister. Rita's parents divorced when she was young, and Rita's mother returned to her natal village. There was a single, poorly maintained very old hut on the property. There was no well, toilet or fence to protect the homestead. Rita had a son out of wedlock in 1980. She sent him to live with her mother. Rita moved to the city where she secured employment as a domestic worker. In 1994 Rita married another man and moved to his village.

---

[51] The sample was taken from a cross-sectional national level sample, 1994 Zimbabwe\Demographic Health |Survey (ZDHS), part of a 13-year project to help developing countries to conduct and analyse surveys on population and health (MACRO International, Inc., 1994). MACRO International, Inc. has been conducting the DHS surveys since 1986, and funding for the DHS surveys has been provided primarily by USAID (United States Agency of International Development). The surveys generally encompass issues of health and population. In general, the DHS surveys aim to get a nationally representative sample of women in reproductive age, and for the 1994 ZDHS the sample includes 6,128 women aged 15-49.

She gave birth to three children and established an independent homestead with the assistance of her husband. Rita farmed in the rural areas while her husband worked in town and sent her some remittances. Rita divorced her husband when he married another woman. She returned to Ndamba, her natal village, with her three minor children and resided in the homestead that had been constructed for her now deceased aunt. The *sabhuku* insisted that she return her children to their father since he was alive. Rita remained in Ndamba as she relied on her natal relatives to assist her in her illness. Rita was frequently too ill to provide for herself and relied on the *sabhuku*'s wife and *sabhuku*'s children to assist her with fetching water, cooking and care. Rita's adult son lost his job during operation Restore Order and moved in with her. The *sabhuku* gave the son some money and advised him to go and find his own people. The son went but failed in his bid to claim his birthright to land and returned to his mother's place.

5.5.3 Netty: divorcee who used the communal area residence as a fallback

Netty's was one of two single women-hearth-holds that maintained a base on a male-headed homestead in the study villages. She was the *sabhuku*'s daughter and 27 years old in 2005. Netty attended primary school (seven years of schooling), after which she secured employment in Harare as a housemaid. She became pregnant, left Harare, and went to live with her divorced father-in-law in the communal area while her husband was employed in the urban area. Netty focused on communal area farming. Netty and her husband divorced shortly after her baby was born. She returned to Ndamba and moved in with her parents. She left the child under the care of her parents and resumed work as a domestic maid. While she was working, she lived with her employer. After two months, she moved to work in a shop in Mutare. She was left after working for two years. She came to Juru Growth Point and secured a job as a shopkeeper through the local ZANU project co-coordinator. She had another baby but did not marry the baby's father. The baby lived with Netty's parents in Ndamba village. Netty helped her parents with expenses for thatching, plastering and glazing their homestead. She rented a room at the growth point. She contributed regularly to the Ndamba village burial society. She was also a recipient of food aid in the village.

Netty's residential pathway illustrates the importance of natal claim to residence as a fallback. She was able to rely on this between work and marriage. Netty's experience also shows single women structure rural-urban connections to manage insecure livelihoods in communal areas and insecure urban employment tenure. While in a marriage, a man migrates and leaves his wife to develop the homestead, a single woman hearth-hold has to rely on other kinship structures to look after her children when she moves on the rural-urban continuum. She lost the marital homestead upon the dissolution of her marriage. She secured work-tied housing and lost that when the employment contract was terminated. The importance of communal area community membership and residence as a fallback for single women is highlighted in this case. In Ndamba, Netty was able to benefit from the community through assistance with childcare and recognition as a member of the community. In return, she had to contribute to the community, which she did through subscribing to the burial society. The differential terms on which Alice, Netty and Rita negotiate access to community membership and residence to Ndamba village illustrate the importance of relationships with people in authority to mediate access.

Research in Chimanimani and Buhera communal lands in Zimbabwe discovered that some women who returned to their natal villages had been allocated land in their own right by the *sabhuku* (Izumi, 2006:42). The focus group discussion with village heads in Mwanza confirmed that women could not inherit land belonging to their parents. However, returned daughters resided with parents in their homesteads and cared for parents in their twilight years in return for shelter. If the parents died before the daughter, she was 'allowed' to use the parents' homestead for the rest of her life. The male relatives retained control of the resource but the single woman had usufruct entitlements which however remained contested.

5.5.4 Ann: widow allocated her own homestead in natal village

Ann was a divorced hearth-hold head that had been allocated a homestead in her own right in one of the study villages. Ann was approximately 80 years old. Ann was allocated a homestead by her father who was the *sabhuku* at the time. The Makuku *sabhuku* was her nephew. Ann had never been to school. She could not remember dates and was not able to relate to historic events like Independence in Zimbabwe. Ann lived with her divorced daughter and divorced son and his two children. Ann lived in a homestead which was well fenced, with an independent well and toilet. There was a chicken coop and a kraal for her livestock. The homestead had one brick and asbestos structure and a thatched hut. There was a kitchen garden. Ann controlled the most land in the village (see Chapter 6). She leased out some of the land to land short villagers in exchange for grain. Ann was a successful farmer who also derived considerable income from remittances and beer brewing. Ann was born in Makuku and left to get married. Initially Ann lived, cooked and farmed with her mother-in-law until she had borne three children. Ann was then allocated her own homestead. Ann's husband was a policeman in the urban area and she was farming in the rural area. It was through the proceeds of the husband's wages that Ann and her husband developed a homestead and she farmed the arable land. The husband died after a short illness.[52] Ann's children were very young. The deceased husband's brother, who had three wives, wanted to marry her in a *kugara nhaka* marriage, but Ann refused the marriage offer and remained on the marital homestead. Without the income from her deceased husband, Ann began to struggle with the homestead maintenance, child care and organising financial resources to fund her farming activities. Ann's father observed how she was struggling to raise her children; he took her back to Makuku. In Makuku, Ann initially resided with her children in her father's homestead. The father, who was the *sabhuku*, later allocated Ann and her children their own homestead and helped her to construct a dwelling and clear the land for farming. Ann financed her activities through proceeds from moulding bricks and working on other people's land. Ann left her homestead in her father's village when she married another man. They had two children after which the man died. Ann returned to her homestead in Makuku. One of her sons married and when divorced moved back to his mother's homestead with his two daughters.

Ann's was the only one of all the 22 case study hearth-holds that had established her own homestead following the end of a marriage. Ann's homestead access was through marriage and later through birth claim in her natal village. In her natal village, she was allocated land in her own right by her father who was the *sabhuku* at the time. This is contrary to findings in another communal area in Zimbabwe that women who returned to their natal villages faced land problems and were reluctantly accepted by their parents (Gaidzanwa, 1994:115). In her natal village, Ann was able to access assistance with raising her children, farming and the establishment of her homestead. Ann's ability to retain her claim to her homestead in her natal village after she remarried illustrates that single women's land claims are not all tenuous. Ann coped with the loss of her husband's wages by returning to her natal village where her father subsidised her and helped her to establish her own homestead. After relying on her father's homestead, Ann successfully moved out and repositioned herself on the rural-urban continuum by selling beer to access financial capital. The latter enabled her to secure agricultural inputs and carry out farming activities.

5.5.5 Cathy: widow trying to re-establish herself after loss of marital home

Cathy is another hearth-hold that was based as a subunit of a male headed homestead. Cathy who was a widow, lived with her two children on her father's homestead. Cathy's father was the *sabhuku*. Cathy was born in 1982 and was 23 years old in 2005. She spent 13 years at school. In 2000, Cathy went to work as a domestic help in

[52] Ann was the eldest person in the two study villages. A combination of illiteracy and advanced age made it difficult to get her to relate to calendars and/or the Zimbabwe independence. It was not possible to extract this information from other villagers in Makuku because as far back as they could recall, Ann was living in the village a single woman. In addition Ann's husband did not die in the village where she was residing at the time of the fieldwork.

Harare and lived with her employers. Cathy left this residence when she got married in 2001. Cathy was married as a second wife after her husband's first wife had failed to produce children. When Cathy married this man, she went to live at the man's homestead in a communal area. The husband was working in town. The marriage produced one son. Cathy's homestead was destroyed by lightning[53] in 2003. She moved back to town and her husband rented accommodation for her away from the first wife who also lived in town. Cathy was not working and her husband provided for her and her child. The husband used to alternate spending a week at Cathy's and another at the first wife's place. In 2003, the husband died of meningitis. Cathy returned to her parent's house in Ndamba with her son. The husband's family said she could return to her natal village, as she was young and needed to work to look after her son. The son lived with Cathy's parents in Ndamba village. Cathy resumed her work as a domestic help after the death of her husband. She worked in Harare for one year but lost the job when her employers emigrated. Cathy lived with her employers during the time she was employed. She secured another position as a domestic worker in Harare in January 2005. On 20 May 2005, Cathy lost her job in Harare because she suffered from incessant headaches. Cathy resided with her parents in Ndamba when she was not employed. She secured another job, which she lost again after she became ill. In September 2005 she got an office job in Harare but, after working for only two days, she had a minor stroke and had to leave the job. She returned to her parents' village.

Cathy's case shows the residential pathway of a hearth-hold that lost her residence claim to the virilocal homestead at the end of marriage. Cathy's marital status, ill health, circumstances of destitution and the responsibility of caring for a dependent young child undermined her ability to negotiate continued access to the marital homestead after the death of her spouse. That both Cathy and Tracy found it relatively easy to regain access to homesteads and rejoin community members in their natal village shows the importance of kinship in facilitating access to homesteads. The natal home was the most secure option as there was the least risk of eviction. That both women were able to regain their natal residence more than twice shows the importance of natal residence as a fallback for vulnerable hearth-holds. As members of the community, the two women benefited from childcare arrangements, assistance from neighbours and support in illness. Cathy's illness increased her vulnerability. The younger woman tried to reduce her vulnerability by securing employment away from the risks of agriculture, but her illness made it impossible for her to work for any length of time. The risk of work-tied residence was realised as the young woman lost her job and residence every time she fell ill. The constantly altering nature of

[53] People believe that witchdoctors are able to make and send lightning (Vijfhuizen, 2002:213). As a result, they shun places that have been struck by lightning fearing that the bolt may be re-sent. Others claim that when lightning strikes a place once, it 'lays its eggs'. As a result, the lightning will strike the same place again.

Cathy's hearth-hold's rural-urban connections the challenges faced by women who have to balance reproductive duties with the need to earn cash income. The illness which constantly interrupts her working life adds another dimension to the life story. The life story vividly illustrates the changing roles of the rural home as a fall back position between marriages/employment opportunities and a repository for young children and eventually the incapacitated adults. This and other cases show the interconnectedness between the rural and urban and the diverse movements of people in response to the various stimuli.

The last five cases have illustrated that hearth-holds left their marital homesteads because of widowhood, divorce, exclusion from the community, witchcraft accusations or the threat of impoverishment. Although the five hearth-holds had different individual circumstances, they all had young children when their marriages ended. This increased their need for support and precipitated their decisions to return to their natal villages. All five cases made different decisions concerning their young children. Ann, Netty and Cathy moved to the natal village with their children. Rita and Tracy left their children in the marital village. Rita was too poor to provide for her own children, and her natal relatives were reluctant to assume responsibility for them. Rita's illness increased her dependency on her relatives and imperilled her capacity to provide for her own children. Tracy was not allowed to take her children with her when she was evicted from her deceased husband's homestead following witchcraft accusations. Netty and Cathy decided to split their hearth-hold residence between the growth point and the communal areas because their conditions of employment and residence were incompatible with child-caring responsibilities.

All the women used the natal village as a fallback as their marital residence had been lost as a result of the demise of the marriage. Three of the five women had been able to establish an independent hearth-hold by occupation of homesteads that belonged to their deceased parents and allocation by the *sabhuku*. The other women obtained residence in their parents' homesteads. Ann's pathway demonstrates that living with parents was in the past a transitory phase for single women hearth-holds. The conditions under which younger single women hearth-holds negotiated residence in Ndamba and Makuku indicates that these women may increasingly be taking up permanent residence in their parents' homesteads. Four of the five women attempted to regain access to residence through remarriage. The strategy was not successful for the women hearth-holds described here and resulted in a temporary displacement of the women hearth-holds from their natal villages. The attempts at remarriage had produced more children. The women risked contracting HIV/AIDS in their attempts to renegotiate marriage. The repeated attempts by four of the five single women to contract a marriage shows the importance and hope that single women place in the relationship. This may mean that these women perceived the potential benefits deriving from marriage to be greater than the risks.

Although all five women returned to their natal villages, the conditions under which they rejoined these communities differed. The hearth-holds who were related to people in decision-making positions, i.e. the *sabhuku,* were welcome to bring their children on their return. The other hearth-holds were not allowed to bring their children into the natal village. This shows that hearth-holds' community membership conditions are not homogeneous. The natal residence was also conditional on the availability of residence and the terms that single women's relatives gave for her residence. The high demand for land on the market, high mortality and dependent children undermined single women's claims to homesteads in their natal villages. The land interests of the women's children were compromised when the hearth-holds left the marital village and relocated to the natal village.

5.6 Hearth-holds that gained homesteads through the informal land market

This section will illustrate how people move into Makuku to take advantage of the land market, through employment or on marriage. In Makuku, there were women whose husbands had migrated into Zimbabwe from the neighbouring countries of Malawi and Mozambique to work as commercial farm workers. As these immigrants did not have a communal home in Zimbabwe to retire to, they negotiated to set up their homesteads in their wife's village of origin. Other people who did not originate in Chikwaka took advantage of the illegal land market to purchase land and establish homesteads in Makuku. The diversity led to variations from the generalised norms about women's experiences on the termination of marriage, as illustrated by the two cases selected from the sample of 22 hearth-holds based on 18 homesteads.

5.6.1 Matty: immigrant widow consolidating a successful homestead

Matty was a widow aged between 65 and 70 years, who lived on a homestead that her and her now deceased husband bought from other villagers in Makuku. After the death of her husband, Matty continued to live on the homestead. She was living in Makuku village in 2005 with two orphaned grandchildren. The house was a three-bedroom brick under tile structure built by her husband when he was alive. The property was well fenced and maintained. There was a good orchard, and a well-maintained thatched kitchen, protected well, cattle pen and chicken coop. Matty lost a 32-year-old widowed daughter, a 26-year-old married daughter and her husband between 2000 and 2005. She had a house in Harare and her only surviving child, a divorced daughter, lived in the United Kingdom. She had one cow on the premises (from the NORAD heifer project, see Section 4.7), two goats and 28 chickens.

Matty came to live here in 1976 after her husband retired from his job in town. The couple purchased the land from the *sabhuku*. The husband had until then been employed as a factory worker in Harare. Matty's husband died after a short illness in 1980. Matty succeeded to her husband's estate as all his male relatives had died before him. She also inherited a house in Harare, a plough, scotch cart and bicycle from the deceased husband's estate. She collected rent from the house in Harare and received remittances from her daughter in the United Kingdom. Matty hired labour for the collection of firewood and the maintenance of her homestead garden. Matty travelled to the United Kingdom for a medical check-up during the research period.

The case illustrates the experience of a person people who obtained homesteads on the market and managed to find a way to ensure access to non-farming income which in Matty's case was the remittances from her daughter in the United Kingdom. This in itself was exceptional. Although this hearth-hold had access to land, in the village, she could not rely on the community to support her with agricultural tasks and homestead maintenance. This made it vulnerable. Matty mediated this by paying for services like fence repair and homestead maintenance on the market. The case highlights the importance of having entitlement to assets in mediating vulnerability.

5.6.2 Betty: ill immigrant widow maintaining a vulnerable hearth-hold

Betty was a widow who lived on a homestead that she and her deceased husband had purchased on the market in Makuku. Betty was 36 years old in 2005. She was a former commercial farm dweller who moved to Makuku with her husband during the 2002 land invasions. Her parents migrated from Malawi and she was born and raised on a commercial farm. Betty had a son when she was 14 years old. She later married a long-distance truck driver and they had three children. Betty and her husband were evicted from the farm during the fast track land reform. They moved in with friends in Makuku where they stayed for about half a year. Later they purchased their own homestead from the Makuku *sabhuku* and constructed a kitchen and a bedroom. The homestead is on the stream bank[54]. The homestead has not been cleared of bush though the residents have been living there for two years. There is no fence, toilet or well. There was a poorly constructed mud and thatch kitchen and a two-roomed brick and asbestos structure. Betty's husband fell ill and died in April 2005. He was suffering from tuberculosis. Betty was also suffering from tuberculosis. Betty lived with her 15-year-old daughter and her three grandchildren whose mother was Betty's daughter. The grandchildren's parents worked and lived on commercial farms. Betty's youngest daughter of 15 was at school. When Betty's kitchen structure collapsed, she had it fixed by her informal liaison. Betty made frequent trips to the commercial farming areas. She was a full-time commercial farm worker. Now that she no longer lived there, she made the daily trips to the farms throughout the planting season. Sometimes she stayed away for up to two weeks. The children begged for food from the other villagers. Betty was arraigned before

[54] Zimbabwean conservation legislation prohibits settlement on the stream bank (Bell and Hotchkiss, 1991:204).

the village court twice and threatened with eviction from the village because of her neglect of the children who lived with her. This was because in her absence, other villagers were obliged to feed the children and provide for them. Betty's 15 year old daughter also worked on commercial farms with her mother during the weekends. Women in Makuku gossiped that Betty was promiscuous.

Betty's case describes the experience of a hearth-hold that is trying to survive in the communal area with little asset entitlement. She has no kin support and limited financial capital. The case shows the limited opportunities for people whose life circumstances force them to rely entirely on the rural economy. Betty's hearth-hold vulnerability was heightened when she lost her husband who had a waged job. The case was compounded by the fact that the husband died before the couple had established the homestead. The homestead was purchased on the market. Unlike Matty, Betty had no cash and therefore needed external assistance on her homestead. She was unable to rely on the host community for assistance with homestead repair, care of her children or any kind of support, as she was marginal as an immigrant. Unlike Cathy and Netty, Betty could not go and work in the urban areas as she had no-one to look after her children. The community recognised her as a resident but she was not entitled to support from them.

5.6.3 Implications

Betty, Matty and Alice's cases explain the circumstances of widows whose husbands entered the village through the market. Betty's experience shows that immigrant hearth-holds try to cope with the challenges of residing in communal areas, raising small children at the same time and having little or no access to financial capital.. Although her husband purchased a homestead on the market, Betty's ability to mobilise community support is limited. Even though some friends assisted them to settle into the village, they could offer only limited support. Betty's inability to care for her children increased the chances of her being evicted from the village and losing the residence. Matty's strategy of paying for services and limiting her reliance on the host community reduced her vulnerability. There were two well-built homesteads in Makuku 3 lying vacant. The former occupants, both women whose deceased spouses had bought land in Makuku, had deserted the homesteads after they lost their spouses. Both women were reported to have returned to their deceased husbands' villages, as they felt isolated in Makuku community.

The residents of homesteads acquired through market transactions are not subject to the patriarchal rules governing the transfer of property upon the death of a woman's husband. However, these market residences are vulnerable in their own way. All three homesteads in this category in the study were developed on land that was transacted illegally. There was always a risk that these might be repossessed by the village head as in the case of Alice, or the widows could be evicted by the State as explained in Chapter 4. This remained a real threat to these women. Matty, Betty and Alice were marginalised by the communities they entered through the market. Alice had to leave the village. Betty's habit of regularly abandoning her children increased the risk of eviction from the village. Matty's economic independence reduced her dependence on the host community and therefore her vulnerability. Alice and Betty were more vulnerable as they were younger and had dependent young children. Betty managed to retain residence of her marital homestead but, unlike Alice, Betty had no natal village to fall back on as her commercial farm residence was lost during the fast track land reform programme. The three immigrant hearth-hold residence pathways show great diversity in the way single women dealt with risks to residence, the loss of their financial capital and the outcomes thereof.

5.7 Single women managing challenges of shelter, ill-health and children

Single women hearth-holds who head homesteads in the community have certain obligations on their homesteads, which they have to meet in their day-to-day living. These are the construction and maintenance of shelter, caring for the ill and young in their homesteads, and looking after their children. The hearth-holds traditionally rely on the community, adult children or their spouses for assistance with these tasks. The extent to which hearth-holds can mobilise community assistance varies with individual capacities, their position in the community and the manner in which they have gained community membership. The hearth-holds also need financial capital to fund the various obligations and pay for certain services. The case studies showed that after the end of the marriage which in most cases ended the migrant labour rural–urban connection, single women hearth-holds had to reposition themselves in order to ensure a continued access to goods and services to enable the to establish and maintain their residence and community membership in the communal areas.

Shelter or housing is important to provide women with cover from the elements, security for their possessions and privacy. In Chapter 4 case 4.6.1 described how a young couple developed a communal area homestead using a combination of family support and financial capital earned by the urban migrant husband. With the exception of Ann, all single women hearth-holds in Chapter 5 lived on homesteads that had been developed by their husband during marriage and/or by their parents. This shows the challenge that single women have to organise residence and develop shelter on their own. One of the reasons is the difficulty for these women of simultaneously earning cash income and maintaining the communal area residence and community membership. A 2006 study on experiences of widows in Zimbabwe confirmed that women struggled with the construction and repair of residences. It was difficult for the women to approach men to assist, as the men's wives did not trust the single women. The study also showed that men demanded sexual favours in return for the assistance (Izumi 2006:35).

In the present study, the homesteads currently occupied by independent single women hearth-holds were constructed by the parents or husband before the women arrived or became single. Only one single women hearth-hold in the sample had constructed their own shelter. This shows that hearth-holds generally had limited access to money, labour and building materials for shelter construction. However, the best constructed and best maintained homesteads in both Ndamba and Makuku villages belonged to elderly widows with sources of cash. This was through remittances, which they obtained from adult children, rental income from an urban property or income from a successful non agricultural venture that was pursued by the hearth-hold head. This included trading, brewing beer or manufacturing clay pots. Leaking roofs, broken or missing fences, decayed buildings and collapsed wells were common problems for single women in both the study villages. Activities like fencing, roofing and maintenance of infrastructure and buildings were male tasks that also require a considerable financial investment. The extent to which single women could rely on the assistance of male relatives for the maintenance of shelter was limited. This was confirmed by *masabhuku* in a focus group discussion, in which they indicated that the withholding of assistance is a common strategy used to 'nudge' women especially those who had refused *kugarwa nhaka* out of the marital community.

Despite the fact all hearth-holds in Ndamba were related to the *sabhuku*, they had to mobilise their own labour, networks and finance for maintenance of housing structures. These included paying for service on the market, relying on adult children or sons-in-law. The hearth-holds that could afford to obtain homestead maintenance services from the market were elderly women with adult children who were employed and sent them remittances Lizzy was unable to do anything about her collapsing shelter as her only child could not afford to remit any money to her, Matty who had access to remittances from a daughter in the United Kingdom and rental income from a house in the urban area relied on the market and hired people to maintain her homestead. Betty an immigrant with neither remittances nor relations in the community entered into informal liaisons with men, and Ann whose father was the *sabhuku* relied on her father's assistance for the maintenance of her shelter. The poorly maintained infrastructure rendered single women's possessions vulnerable to theft and livestock. Betty and Rita's cases demonstrate the vulnerability of hearth-holds who cannot mobilise labour and financial capital to maintain their shelter. Vulnerability to theft was experienced differently by single women in Ndamba and Makuku. Tracy and Mary in Ndamba relied on neighbours and kin to provide security for their residences, whereas in Makuku Lizzy's home was ransacked in her absence. Only four out of the 12 single women-headed homesteads had a well and toilet. The rest did not have any of the facilities. They either had never been constructed or had collapsed due to lack of maintenance. The women relied on the facilities of neighbours or the bush. The hearth-holds which co-resided as a sub-unit of a larger family group or lived in employer-tied accommodation were not responsible for the maintenance of their shelter.

Illness is one of the major determinants of single women's residence location. Widows have reported poor eyesight, ill health, physical handicaps. Poor health contributes to their inability to support themselves or others (Folta and Deck 1987:332). Ill women rely on the community for support. In the present study, the focus group discussions revealed that natal kin were responsible for the care of ill adult women, irrespective of their marital status. This was indicated by both the presence of ill adult daughters in their natal homesteads and the reported incidence of adult daughters who died under the care of their natal family. The experiences of Tracy, Cathy and Rita demonstrate the importance of family support to single women in times of illness. Single women with access to money can seek medical assistance in the private sector, as shown by Matty and Mary's experiences. Alice's experience shows that churches could offer some assistance to destitute people, but in Alice's case this assistance was conditional on Alice abiding by the rules of the church, which prohibited her from visiting the clinic. Alice lost the church's support when she took her son to the clinic when he fell ill.

Children require care and nurturing when they are young. They cost money in terms of schooling and care. The birth of children enhances a married woman's status. In a marriage, children are the responsibility of both parents. Although children increase a woman's chances of retaining her marital homestead in widowhood, they can be a liability when young. In the present case-studies, the women who had given birth to children had difficult decisions to make on the demise of their marriage. The hearth-holds either remained in the marital homestead or returned to their natal communities. Rita, Lizzy and Ginny's experiences showed that the contraction of wage earning opportunities following structural adjustment and the demise of the economy meant that adult males also failed to secure employment as these women's children remained unemployed. As a result, these hearth-holds did not receive remittances from their sons who nevertheless still migrated to the urban areas in search of work. In addition, the rural homesteads became more important as a fall back and primary residence for these men who relied on their mothers to look after them. In some instances, these men cost the single hearth-holds income as they resided in the urban properties from which the single women who owned urban properties could have obtained a rental income to offset the failure of remittances.

Children raised among their patri-kin are entitled to inherit land on marriage and/or through inheritance. The children's patriarchal relatives have a customary obligation to raise them. The ability of a widowed woman to mobilise childcare support and assistance is conditional. Anna was expected to marry the brother of her deceased husband, failing which her husband's relatives withheld assistance. Cathy, Ann and Jane decided to leave their deceased husband's villages with their young children. The experiences of Cathy, Jane, Rita and Ann's children show that children who leave their patriarchal village lose their rights to homesteads. The children had a difficult if not impossible task of reclaiming their residence and community membership in the paternal village when they became adults. The experience of Rita's son, who was unable to claim his rights to land in his father's village and had to return to his mother's natal village vividly illustrates this. Hearth-holds with children have to decide between staying in the marital community or reducing their vulnerability by leaving the marital community and returning to their natal village where they would have better access to social security. The latter option is chosen, however, at the expense of their children's homestead rights in the children's patriarchal community. One of the strategies formerly married women use is to decide to remain in the marital village under very difficult circumstances. This was a strategy used by Lizzy.

The experiences of the women who brought their children to their natal villages illustrate that this too has limits. Women who managed to bring their children to the natal home were engaged in constant fights with their relatives who were unwilling to assist them when their children were ill or needed school fees. Single women in the sample who were in the natal homesteads were worried about the future of their children because the woman's natal families threatened to evict the children upon the death of the mother. The situation was worse when the women lacked independent means and were not well, as illustrated in the case of Rita in Ndamba village. Rita had to leave her children in their father's village or risk losing her right of residence in Ndamba. Ginny on the other hand successfully resisted attempts by the Ndamba *sabhuku* and community to evict her ill son who had followed her to the village.

A hearth-hold's natal relatives resent their daughters' children because the children pose a threat to the long-term residence interests of the people in their mother's natal village because, as these children become adults, they form families and in turn negotiate access to homesteads, residence and community membership. This increases competition for scarce land, natural resources and the communal area land market. Hearth-holds do not have similar experiences even within the same village. In Ndamba, for example, Cathy and Netty's children were welcomed whereas Rita's were not, because of their different positions in relation to the *sabhuku*. Rita was ill herself, had no means or a living female relative to assist her with childcare. The Ndamba *sabhuku*'s attitude towards Ginny and Rita's children exemplifies how the community excludes people in order to limit increasing demands on the social security system of the village.

The single women residing in Makuku on a natal basis faced similar resentment, not at the level of the *sabhuku* but from individual families. As a result, experiences were widely divergent. However, Betty, an immigrant in Makuku, was hauled before the village court and threatened with eviction as she constantly failed to take care of her young children. Her behaviour placed demands on the community, as other people had to look after the children. There were some young men in Makuku who had returned to the village with their mothers as young children. When these men became adults, the Makuku *sabhuku* and community denied them customary land given to married men, as they were not viewed as full members of the community. Some of these male children purchased land on the market in an effort to secure their claims. Even then, these children were regarded as squatters, along with the former commercial farm workers, by those who considered themselves as the rightful occupants of Makuku. These children faced a real risk of displacement, and tried to mediate their vulnerability by negotiating access through political activism and entering into informal relationships which they hoped would evolve into marriage.

All the single women hearth-holds in the study accessed residence conditionally. They constantly had to negotiate residence and community membership in the marital, natal or market-secured community. The hearth-holds resisted eviction, marriage, widow inheritance and attempts by the village head to stop them from looking after their children. The different decisions that widows took on the position of remarrying the husband's brother or the location of their children illustrates that widows like these have more room for manoeuvre than previously believed (Hindin, 2002). The examples of single women moving from one residence to another between marriages, work and childbearing, and the reasoning behind each move, demonstrate how women used the networks, financial resources and own labour to negotiate residence in the communal areas.

5.8 Conclusion

It is important for single women to secure communal area residence as it is from there that the women can organise their livelihoods, access social security and have the opportunity to mediate their vulnerability. Although some of the livelihood activities are located outside the communal areas, access to a communal area homestead remains important, as it continues to be relatively cheaper and more secure than the alternatives of urban areas, commercial farming areas and/or the growth point. Having residential land means access to a community as well as a place to live. Single women hearth-holds gain access to residence in communal areas in a very uncertain environment. Women are bounced back and forth between kin and affines and have a very unclear and insecure status with respect to residence. This is not inconsistent with the fact that they have access to residential land, but this access is contingent upon the sets of conditions that have been identified.

Homestead access is subject to constant renegotiation because of the unstable environment within which hearth-holds negotiate access. The multiplicity of variables that influence single women's residence access is inconsistent with findings that infer that customary tenure systems generally result in the eviction of widows and restrict hearth-holds' access to residence in their natal homestead (Andersson, 1999; Huisman, 2005; Goebel, 2005). The findings of this study are that single women have more room for manoeuvre in deciding where to locate their residence. Hearth-holds deserted homesteads, were nudged out by the community or were pulled out by their natal relatives. These tendencies are neither the dominant nor the only outcomes of single women's residence status following the change of marital status. There were no cases in this study of hearth-holds that were violently evicted from communal areas.

The rural-urban connections were important in determining single women hearth-holds' access to community and residence in the communal areas. In this chapter the historical importance of the wage labour contracted in financing and the construction and maintenance of a homestead was highlighted. It emerged that all of the homesteads that single women hearth-holds occupied were constructed through migrant earnings by a husband, brother or other male kin. That the hearth-holds heading and managing their own homesteads struggled to maintain and or improve these homesteads confirmed this observation. Nevertheless, the hearth-holds continued to try and organise their livelihoods across the spatial boundaries in order to maintain their residency. This involved moving between the rural and urban areas and splitting the hearth-holds between the rural and growth points or urban areas. This happened when the hearth-hold head secured employment and had to leave their children in the rural areas. Hearth-holds' mobility was also informed by the need to balance gendered responsibilities of raising children and caring for the ill. As a result, it is not possible to generalise the direction, frequency or consequence of rural-urban connections for single women hearth-holds. However, the chapter illustrated that these connections were important and the communal areas were inextricably linked to other spatial enclaves in the country.

Hearth-holds accessed residence through kinship relations, marriage and the market. Employment-tied residence was limited to the duration of the work contract. The hearth-holds that accessed homesteads on the market remained marginalised from the community and had limited access to social support systems in the community. Access through kinship ties emerged as the most accessible to single women as they had limited finances and access to employment opportunities to negotiate homesteads on the market. The kinship, homestead and community membership were earned through marriage and birth. Change in marital status wrought by death or divorce altered the conditions upon which the woman maintained residential and community access. The resultant change in terms of access meant that women had to resume negotiations with different people who assumed control of the homestead and community membership. The options open for the women were mediated by the interests of those controlling the land and community. These included the village head, community and kin. The flexibility of communal area land governance gave rise to a variety of possible outcomes for single women's residence access.

There was no single outcome of these negotiations as each single woman's situation was unique. Women born on the commercial farms or whose natal kin were all deceased did not have the option of returning to the natal residence. Hearth-holds returned to the natal village if they perceived the benefits of residing there to be greater than those of remaining in the marital village. Single women hearth-holds left the marital village when they failed to mobilise labour to construct and maintain a homestead, were accused of witchcraft, were ill, were of a young age, faced the threat of impoverishment caused by the loss of a wage-earning spouse and had to tackle the need for the support associated with raising young children. These experiences underline the importance of the natal family as insurance for women in their illness, old age or when nurturing of children. The natal residence was also conditional on the availability of residence and the terms that single women's relatives imposed on her residence. The market, high mortality and dependent children undermined single women's claims to homesteads in their natal villages. The kinship system is under siege from the increasing number of single women and divergent ways of gaining residence in the communal areas. This has resulted in loss of support and increased exclusion and marginalisation of single women who have increased their dependence on their immediate families; it has also resulted in new outcomes for women who gained land through the market and employment as they escaped the negative sanctions of the kinship system normally imposed on widows in a patrilocal marriage. The precarious positions and residence deterioration of single women hearth-holds without active rural-urban connection underlined the importance of these linkages to the hearth-holds' residence and livelihood in the communal area.

The vulnerability of single women hearth-holds' residence in communal areas is linked to the vulnerability of their children's residence rights. Children have the most secure residency claims in their paternal village. These claims are normally safeguarded by the mother's continued stay in the village. The destabilisation of marriage undermines a hearth-hold's living environment, sometimes leading to its abandoning the marital residence. This compromises the children's paternal land claims in the long term. There is a catch-22 situation for the women: whether to decide on staying and increase their children's opportunities to gain village membership or leave and potentially risk their children's losing their paternal residency claims.

The importance of support and social relations for single women cannot be overstated. Villages like Ndamba, which was a relatively cohesive community, are in a better position to assist desperate single women. However, access to this community was increasingly restricted. This is the reason why single women had different experiences in accessing this support. The support was mainly extended to kin and single women born in the village. In Makuku, there was less restriction on entry into the community. However, it was harder for single women to mobilise support because of the diversity introduced into the community by the market. There were limits to the kind of support that single women, even among their own kin, could draw from the community. Market access remained most difficult for single women and was even more difficult to maintain because of exclusion by the host community and the isolation that single women experienced. This chapter has confirmed the insecurity of residence obtained on the market as it is subject to repossession by the *sabhuku*. This finding is consistent with those of Chimhowu and Woodhouse, (2006b:36). Single women whose hearth-holds had entered the community through the market coped with community isolation by forming liaisons with men, depending on other migrants or obtaining services on the market.

There is no general impact of the various factors on single women's residence. The conditions may interlock to produce extreme vulnerability, and the reverse is also true. The most vulnerable single women in this study included the ill, those with dependent children and those lacking access to reliable, regular income. The women who had inherited marital residence and had adult children who had successfully migrated or had managed to organise another consistent source of income were the least vulnerable. However, the residences remained subject to constant negotiation.

Chapter 6

A field not quite of her own: single women's access to agricultural land[55]

6.1 Introduction

Chapter 3 has shown that demand for land in communal areas has been increasing because of increased population density caused by the displacement of people from urban areas and commercial farming areas. It also established a counter tendency that population increase, resettlement and HIV/AIDS have resulted in the spatial displacement of people from their land in communal areas, in response to their increasingly insecure agrarian livelihoods. Hearth-holds displaced by the breakdown of their marriage, illness or the termination of their employment contract, vacated their residences and arable land in their place of origin and had to negotiate for new residence and/or land access in their new destination. This chapter focuses on how indeterminate governance institutions, mobility, high morbidity, mortality and rural-urban connections intersect to produce various opportunities and constraints for livelihood resource access for single women and other marginalised people in the communal areas. The focus on hearth-holds and single women's land access exposed various 'extra-legal' methods, which single women used to access land, residence and community to reduce their vulnerability in communal areas. The chapter is organised in four sections. The following section deconstructs the meaning of 'vacant land' in the context of the communal areas. It defines the concept of *matongo* and shows how mobility, mortality and legal ambiguity identified in Chapter 3 nurtured *matongo*. The second part shows how single women gain resource access through the *matongo* layer and identifies the vulnerabilities of these opportunities. The third part focuses on the impact of the governance context on *matongo*. In the final section, the implications of the findings are discussed and conclusions are drawn. The existence of *matongo* reflects the vulnerabilities that people in communal areas confront, but it also creates new opportunities for them.

6.2 Unravelling *matongo*

Historically, rural-urban connections consisted of cyclical migratory movements of mainly men between rural and urban areas. The men earned a wage some of which they remitted to the communal areas. This formed the main source of financial capital for communal area livelihoods. The nature of the connection has changed over the years because of the war, independence, structural adjustment, economic

[55] Submited to World Development Journal.

crisis and operation Restore Order. People move between the rural and urban areas in response to the various stimuli. The variety of stimuli of these movements means that it is not possible to generalise about the nature, direction, volume and frequency of these movements. High mobility has resulted in people having an indeterminate residential and occupational profile. This mobility is accompanied by an enduring availability of vacant land in the communal landscape. This vacant land is described by the Shona term of *matongo* (singular *dongo*), which refers to previously inhabited places (Andersson, 1999:559).

The existence and importance of *matongo* became clear through the compilation of a vacant land register during the fieldwork undertaken for this study (see Chapter 2). Table 6.1 shows an excerpt from the Makuku vacant land register compiled in 2005. The table's second and third columns identify the type of land and the approximate period for which the land had been *matongo*. Both arable and residential land can be *matongo*. That means that when the person considered to be the rightful owner of the property permanently or temporarily stops using the land, then it is a *dongo*. The fourth column identifies the processes leading to the vacant land, and the processes that led to the land vacancy. These include mobility, mortality, marital dissolution and household consolidation[56] The fifth column identifies the source of the land, which became a *dongo*. The column shows that the land was either allocated on marriage, inherited or even purchased on the illegal land market. This means any category of land can become a *dongo* in communal areas. Although this work focuses mainly on single women, the table illustrates that both men and women have lost and gained land access through *matongo*. The institution of *matongo* creates both opportunities and vulnerabilities for people in communal areas. The sixth column indicates the use and/or user of the land in 2005. The abandoned *matongo* in column 6 on the table describes arable and/or homesteads that were not in active use during the fieldwork period. This is to distinguish these homesteads and arable fields from those that in the absence of the rightful owner were being put to use by relatives, tenants or employees.

Matongo are areas that were used for human habitation in the past and were subsequently abandoned by the owner. As a result, even homesteads and agricultural fields that were put to use by somebody who was not considered by the community to be the one supposed to be on the land, were referred to as *matongo*. The various informants that were used to define *matongo* were community leaders, members of Zimbabwe Widows and Orphans Trust (ZWOT), key informants and single women. Field observation also yielded information, which was helpful in defining *matongo*.

[56] Household consolidation occurs when the residents of two or more homesteads live together. This occurred in the research area when orphaned or widowed people moved in with a grandmother or sister who lived on a different homestead. The homestead that has been abandoned in the process becomes a *dongo*.

The land identified as *matongo* consisted of homesteads that were considered as undeveloped, abandoned, vacant, fallowed arable land and land occupied by single women. The classifications are described in more detail in Table 6.2. The figures used in this chapter are based on homestead *matongo*. The phenomenon of *matongo* was constantly changing because of the fluid nature of processes that gave rise to its existence. As a result, figures are used mainly to indicate the extent of the phenomenon. The figures presented in this chapter were accurate in November 2005. However, the situation will have changed considerably since then.

Tables 6.1 and 6.2 show *matongo* to be a process and not a state. It is a negotiated land access procedure during which various overlapping claims to the piece of land exist. For example, plot 32 on Table 6.1 is a *dongo* because the couple who purchased the land and developed the homestead do not live on it. It is referred to as a *dongo* even though the place is occupied by the couple's employee. There is no clear definition or identification of parameters for defining *matongo*. Table 6.1 was drawn up after compilation of the vacant land register, focus group discussions with *masabhuku* and widows, and key informant interviews who physically identified the *matongo*. Table 6.1 indicates that *matongo* can be defined by use (abandoned, uninhabited), gender (used by single women hearth-hold, time, activity (or lack of e.g. land under fallow), relation to other users, or in terms of normative access frameworks. The categories are not mutually exclusive. The explanations of the criteria indicate the subjective and fluid nature of the concept. The table illustrates the importance of qualitative data in customary tenure analysis. The categories of vacant land are defined in line with the processes that gave rise to them. These are explained in Table 6.2. The village head in Makuku identified the undeveloped, abandoned, absentee-owned land and land occupied by single women as *matongo*. The Ndamba village head identified the abandoned, absentee-owned and vacant land, and places occupied by single women as *matongo*, because single women's occupancy of the *matongo* was considered temporary and conditional. The single women identified abandoned and unused arable land as *matongo*. The single women did not refer to the homesteads that they were occupying as *matongo* Instead, they referred to the places that they occupied as belonging to their deceased father, spouse, out-migrated owner or whatever the case was. I quizzed a single woman hearth-hold on this and she responded that her entitlement to using the homestead made her an eligible user. This disagreement on what constitutes *matongo* indicates that *matongo* is a contested resource. This confirms an assertion that the description of fallow land and land of absentee owners as vacant is a form of contestation (Shipton and Goheen, 1992:309).

The experiences of delineating *matongo* partially explain the logistical problems described in Chapter 2 when I attempted to do a 'complete' inventory of land ownership in Makuku village. Different people move between locations in different response to the multiple contextual factors affecting the communal area environment. Consequently, the timing of a survey is important in determining whether a land

Table 6.1. Register of land held by absentee owners in Makuku village in 2005.

Plot No.	Type of land (residence or arable)	Number of years the land has been vacant	Registered owner	How the land was acquired by registered owner	Use to which the land was put in 2005
1	Both	3	Migrant man	Inherited from grand parents	Single woman*
2	Both	+10	Migrant heirs	Inherited	Single woman*
3	Both	6	Deceased man whose wife left*	Allocated on marriage	Parents farm
4	Both	6	Widow resides in another *dongo*	Allocated on marriage	Single woman*
5	Both	4	Deceased man	Allocated on marriage*	Deceased man's mother farms
6	Both	3	Deceased man	Allocated on marriage*	Single woman*
7	Both	+10	Migrant son	Inherited	Abandoned not inhabited
8	Both	+4	Resettled man	Inherited from deceased brother	Abandoned, not inhabited
9	Res	3	Resettled man	Purchased	Employee growing food
10	Both	10	Migrant children	Inherited	Abandoned, not inhabited
11	Res	2	Man living with girlfriend*	Inherited it from his deceased sister* who had purchased it	Abandoned, not inhabited
12	Res	2	Resettled man	Inherited	Abandoned, not inhabited
13	Both	5	Deceased man	Allocated on marriage*	Nephew farms
14	Res	2	Migrant*	Purchased	Abandoned, not inhabited
15	Both	1	Migrant	Allocated	Growing food
16	Both	3	Returned divorcee*	Use after decease of parents	Agriculture
17	Both	10	Young children	Inherited from grandfather	Single woman*
18	Res	6	Migrant*	Purchased	Farming only
19	Both	5	Grandsons migrated	Inherited from grandparents	Parents farm
20	Both	5	Deceased man*	Inherited from grandfather	Abandoned, not inhabited

Plot No.	Type of land (residence or arable)	Number of years the land has been vacant	Registered owner	How the land was acquired by registered owner	Use to which the land was put in 2005
21	Res	Unknown	Migrant	Purchased	Single woman*
22	Both	5	Deceased man	Allocated on marriage*	Used by the men's mother
23	Both	4	Deceased man	Allocated on marriage*	
24	Res	5	Migrant	Purchased	Employee*
25	Res	8	Migrant	Purchased	Brother farms
26	Both	15	Migrant	Inherited	Abandoned, not inhabited
27	Res	4	Woman who married and moved to commercial farm*	Purchased	Brother farms
28	Both	+20	Returned daughter who is alive and lives in the village*	Assumed control after father's death	Leases some to landless, uses some, rest abandoned
29	Both	5		Assumed control after brother's death	
30	Both	3		Assumed control after migration of son	
31	Res	8	Widow left*	Purchased	Abandoned, not inhabited
32	Res	3	Migrant couple	Purchased	Worked by employee
33	Res	2	Widower moved in with his widowed mother*	Inherited	Nothing, not inhabited

*Indicates single woman access.

Source: Field work 2004/5.

Table 6.2. Criteria for delineating matongo.

Criterion	Explanation
Undeveloped	This is land that has been demarcated and allocated to somebody for the development of a homestead but that has not yet been developed. There are no structures on the homestead. These situations arise when the person allocated land by the *sabhuku* is absent from the village, has taken up residence in another dwelling available to him, e.g. late parents' homestead, consolidates residence with an existing parent or lacks the building materials, labour and money to develop an independent homestead as described in Chapters 4 and 5.
Abandoned (see Figure 6.1)	These homesteads have shelter structures constructed on them but they are neglected and unoccupied. Such *matongo* arise due to death and/or out-migration of the owner. An example is when widows abandon homesteads following the deaths of their spouses, as illustrated in Chapter 5.
Absentee owned	These are homesteads on which dwelling structures have been developed but are not in use by the 'owner'. Usually other people, including employees and relatives, use the homestead. The owner is deceased or has migrated.
Single women occupied	These are a subgroup of the absentee-owned *matongo* and are occupied by single women, because single women do not have primary rights to the homesteads but instead occupy them on behalf of either their children, brothers, or other relatives.

Source: Fieldwork November 2005.

Figure 6.1. Abandoned homestead dongo.

is classified as a *dongo* or not. The link between mobility and *matongo* means that mobility patterns over the year influence the landscape and the classification of land. Given that mobility is a permanent feature of life in communal farm areas, it is not possible to choose a time that would yield a 'true figure in the Zimbabwe context' (Potts and Mutambirwa, 1990:685).

In Chapter 1 a homestead was defined as a piece of land that has been allocated to an individual and developed through the construction of shelter for human habitation. A homestead becomes a *dongo* homestead when it is no longer occupied by the person that the community considers being the rightful owner or is abandoned. Makuku village had a total of 90 homesteads in September 2005. This was up from 68 homesteads counted in 2004. According to the various criteria in Table 6.2, there were 56 *matongo* homesteads in Makuku village in September 2005[57]. Ndamba village with 22 occupied homesteads had eight *matongo*. The difference in the incidence of *matongo* can be explained by the age of the village, the number of homesteads, the governance context and the timing of the survey. The legal uncertainty and plurality that characterise communal area resource governance have given rise to the divergence in the interpretation of *matongo* by the informants. This is reflected in Table 6.1.

Matongo develop for a variety of reasons, including mortality, mobility, traditional beliefs, speculation, government policy, lack of urban security of tenure (Mbiba, 2001; Potts, 1995) and the diverse ways in which people in Zimbabwe manage their rural-urban connections. The re-organisation of settlements in the communal areas under the Native Land Husbandry Act of 1951 gave rise to *matongo*. When the Government moved people from their original settlements into organised villages, the people retained their original homesteads as *matongo*. Andersson's study of Buhera communal in 1999 established that people were referring to that 1951 legislation to justify their claims to a contested space in a communal area (Andersson, 1991). Table 6.1 shows that the *matongo* observed in the research area originated from market transactions of land, customary allocations and inheritance.

Approximately a third of the *matongo* in Makuku village were obtained through market transactions. The people who purchased the land were resident in places where they were employed. These are represented by *matongo* numbers 9, 14, 18, 21, 24, 25, 27, 31 and 32 in Table 6.1. The market-related *matongo* occurred only in Makuku and not in Ndamba because there were no market transactions with land in

[57] The idea to investigate *matongo* developed during the fieldwork. As a result, during the initial survey in 2004, the number of homestead *matongo* in Makuku was not captured. In Ndamba, it was easier to recall the figure as it had remained more or less constant and the village had fewer homesteads which made it easier to take note of mobility and occupation patterns on various homestead throughout the fieldwork period.

Ndamba. The people who purchased land but did not reside in the village were away working on the mines, commercial farmers and urban areas. They purchased and developed land in Makuku so that they could retire to it at the end of their working life and or when they became too ill and or unemployed. These arrangements were a way of maintaining and managing rural-urban connections and resource flows.

The Chikwaka ward *matongo* statistics[58] show that the period for which a *dongo* was in existence in the ward ranged from two months to more than 20 years. These years of vacancy were estimates made by key informants who included village health workers, the elderly and *masabhuku*. A third of the *matongo* had been in existence for at least fifteen years. The 1-15-year-old *matongo* accounted for two thirds of the *matongo* in the ward. Of these, the 0-5-year-old *matongo* made up more than half of the *matongo*. This is associated with the high mortality experienced in the last ten years due to the prevalence of HIV/AIDS and the high mobility of people in response to the changing rural-urban connections. A few of the identified *matongo* could not be dated, as the key informants were not able to give estimates.

6.3 Processes leading to *matongo*

6.3.1 Mortality

The HIV/AIDS pandemic caused a sudden increase in mortality rates in Zimbabwe (Feeney, 2001). A focus group discussion with *masabhuku* concluded that the mortality-related *matongo* had significantly increased in recent years in the study ward. The majority of those dying were in the 18-45-year-old age group. Death gives rise to a *dongo* by depriving a homestead of a household head. This happens when the male head (as the case of Alice) or both spouses die. Chapter 5 discussed how the death of a male head triggered subsequent out-migration of the female spouse to her natal home, growth point or urban areas. The case studies showed that this could be temporary or permanent. It is not possible to generalise. Discussions with some widows that had left their marital homes following the deaths of their spouses showed that the widows' abandoned homestead remained as *matongo* and the widows and/or their children could always return and claim the homesteads. Section 5.6.3 identified two women who had relocated to Makuku when their respective husbands purchased land and developed homesteads there. These people's homesteads in their villages of origin remained standing as *matongo*. When their respective spouses died, the women returned to their *matongo*. The death of a female spouse in a young couple's homestead may also trigger out-migration. For example a widower abandoned his homestead following the death of his wife and returned to his parental homestead, as illustrated by the history of plot 33 in Table 6.1. Chapter 4 identified

[58] Gathered from a small survey with village health workers.

cases of young men who failed to marry and establish independent homesteads and some who married were allocated land to develop homestead but lacked the financial resources to secure labour and material to construct their housing, wells and livestock shelters. These men remained on their parents' homestead while their undeveloped homestead land remained as *matongo*. The process of inheritance ensured that land was passed from the dead to the living. If the heir was too young to inherit a homestead, living away from the village or not yet married, they could not occupy the land and/or homestead that they were entitled to. This increased the incidence of *matongo*.

A new phenomenon of 'upward inheritance' has been observed in cases where adult children die before their parents. The normal inheritance patterns are downward in that descendents inherit the assets. The new process is termed upward inheritance because it reverses the generational order of inheritance. In the current study, parents assumed primary control of the homesteads in a number of cases after the death of their child. Plots 5, 22 and 23 in Table 6.1 are examples of this. The woman living on plots 22 and 23 in Table 6.1 lost six adult children over a period of ten years. She assumed control of two homesteads that had been allocated to, and developed by, her two deceased sons. The deceased daughters had been married at the time of death. The deceased women may have left *matongo* in their marital villages. Such *matongo* are referred to as abandoned *matongo* in Table 6.2.

Traditional beliefs also perpetuate *matongo*. Tables 6.1 and 6.2 show that some *matongo* were linked to the non performance of inheritance transfers ceremonies. Customary practices, normally performed a year after burial, include the distribution of the deceased's moveable and immovable assets (May, 1983; Vijfhuizen, 2002). Some of the families in the current study failed to conduct the ceremonies due to either lack of money or disputes between the interested parties. The affected homesteads remained vacant. Social barriers prohibited the occupation of homesteads by certain categories of people in the study area. An example is the prohibition of the permanent occupation of a parents' homestead by an adult daughter and her children as these are reserved exclusively for a son's wife or a grandson's spouse. This did not preclude the daughters from temporarily occupying the place until the eligible people needed a homestead. The period that the daughter was allowed to occupy the homestead varied but was usually determined by the presence or absence of the 'rightful occupant'.

Another practice occurring in the research site stipulated that, when a man married a wife, he was obliged to build her a kitchen. If the wife died before the husband and he decided to remarry, he had to provide his new wife with her own kitchen as happened in Alice's case. The deceased wife's kitchen was reserved for her daughter-in-law or her grandchildren. Where land was available, the man relocated to another place in the village with his new wife, leaving the previous place as his *dongo*.

People believe that witchdoctors are able to make and send lightning (Vijfhuizen, 2002:213). There was a village in the study ward where three adjacent homesteads became *matongo* after their occupants were killed by a single bolt of lightning. The homesteads remained *matongo* as people were afraid to occupy them because of traditional beliefs that associated lighting with witchcraft and the supernatural.

Historically, people have buried their dead on the homesteads in the communal area. Such places become symbolic in some rituals and descendents are obliged to maintain such land in their family. It therefore becomes very difficult to sell such a place. As a result, such homesteads become *matongo* if there is no immediate kin available to occupy them. The Government of Zimbabwe made several announcements at the monthly ward meeting aimed at discouraging the practise of burying people on homesteads. However, the government lacks the manpower to enforce the decision. This observation was also made by Andersson in his study in Buhera where he concluded that '...state institutions responsible for land issues have a limited understanding of and exercise little control over land issues...' (1999: 553)

6.3.2 Mobility

In chapter three, the historical analysis of the development of Zimbabwe showed how the establishment of rural to urban areas created socio-economic spaces that straddled the rural-urban divide. Consequently, people's livelihoods are organised across the spatial boundaries. The people and goods move between the rural and urban areas to maximise opportunities. This mobility leaves vacant spaces that present opportunities for other people. The rural land holding status of the mobile people in communal areas is subject to debate. There are those who believe that the land holding rights cannot be lost on migration (Pankhurst, 1991:612). Others argue that *masabhuku* have a right to reclaim and re-allocate land that has been left vacant by the mobile people (Berkvens, 1997). The *masabhuku* in Mwanza were ambivalent on the matter because of the multiple interpretations of the legal system, specific local contexts and the existence of plural legal systems in the communal areas. The legal uncertainty increases the vulnerability and decision-making context of the weak, such as single women, young men and those who have purchased land on the informal land market.

Previously these people retained their communal area land claims as they offered social and economic security for urban dwellers in an insecure environment (Potts, 2000a:813). This has also increased the number of *matongo*. The changes in the rural-urban connections and mobility patterns induced by the war, structural adjustment, independence, land resettlement and HIV/AIDS has increased the incidence of *matongo* because women and children were also displaced as whole families ceased the physical occupation of their communal area homesteads. The land redistribution programmes in the 1980s (see Section 3.2.1) created opportunities for people to move to former commercial farming areas. Some of the people retained their communal area land interests, and this gave rise to *matongo*. This is illustrated by plots 8, 9 and 12 in Table 6.1. The change in laws governing ownership of urban property by Africans at independence and increased urban housing security, have resulted in more permanent settlement of people in urban areas. The people who found security in urban areas reduced their reliance on rural land holdings as security. As a result, such people's homesteads remained derelict and abandoned *matongo*.

People who move out of a communal area choose to hold on to their land rights in their area of origin because of the insecurity of urban livelihoods, their wish to be buried at their home of origin and the fact that there is no obligation for migrating people to give up their land rights (Potts, 2000a:809). As a result, it becomes impossible for the market to reallocate the land in an efficient manner i.e. to those who need it most. The people's historical mistrust of government and the legacy of the forced removals during the colonial era and more recently operation Restore Order is another reason migrants hold on to their land in communal areas. Survey findings from the fast track resettlement areas concluded that peasants mistrusted the State and that this mistrust was based on experience. As a result, those who benefited from the fast track land redistribution effort maintained their rights to communal land (Sithole *et al.*, 2003). In August 2005, for example, 2,000 families who had moved onto a former commercial farming area in Goromonzi were evicted to pave the way for the political elites[59]. Another report cited clashes between more than 180 resettled farmers and the political elite who sought to evict them[60]. Operation Restore Order resulted in the eviction of people from the urban areas. A post-operation Restore Order survey reported that 53% of the displaced people relocated to a rural area (Bratton and Masunungure, 2006:16). Processes like these strengthen people's belief in the need to maintain a rural home and confirm the continuity between the rural and urban spaces in Zimbabwe.

Chapter 4 showed that people moved into Makuku village and purchased land on the market to escape land pressure in their own communal areas, to access the river bed for gold panning and to be nearer Harare and surrounding commercial farms-oasis of non agricultural income earning opportunities. Some of these people who buy land in Makuku but work and live elsewhere for example in Harare or commercial farms maintained land claims in Makuku as a fallback and/or final burial place. While they remain unoccupied, these people's homesteads added to the stock of *matongo*. Chapter 5 also showed that some single women abandoned their homesteads after

[59] Financial Gazette, 5 July 2005, p15.
[60] Daily News, 23 October 2005, p4.

they lost their spouse, as they felt isolated. These homesteads also contributed to *matongo*. Chapter 5 illustrated how the increasing break-up of marriages and subsequent relocation of hearth-holds contributed to migration-induced *matongo*. These *matongo* are emphasised with an asterisk (*) in Table 6.1.

There are also *matongo* that belong to people who have emigrated that also contributed to *matongo*. There was a homestead whose owner was reported to be working for the Zimbabwean embassy in Asia. In Makuku, related younger and older people living alone on their respective homesteads consolidated residential units to maximise efficiency of resource use and this increased the *matongo*. Instead of a widowed son and his widowed mother cooking a respective pot of food on their individual homestead, the two domestic units occupied a single homestead and pooled the financial and labour resources available. Examples of consolidated households included divorced or widowed sons who had abandoned their marital homesteads and relocated to their parents'(usually widowed mother's) place, grandparents living with grandchildren and adult sons, some of them married, who lived with their widowed or divorced mothers. This is described in the example below.

6.3.3 Case 6.1: overlapping claims on a *dongo*

This homestead in Makuku was a *dongo* previously occupied by a now deceased senior inhabitant of the village. The *dongo* was located adjacent to the homesteads of three single women in the village. The women were a daughter-in-law, daughter and sister of the late owner, respectively. Their relationship with the late owner entitled them to use the *dongo*. The three women normally used the *dongo* for growing their crops during the rainy season. Alan was a grandson of the deceased *dongo* owner and also related to the three women. This also entitled him to the *dongo*. When Alan married, the women made way for him and his wife to set up their homestead on the *dongo*. Alan constructed a house, dug a well and moved from his widowed mother's homestead onto the *dongo* with his wife. Alan's wife became ill and died after three years. Alan moved back to his mother's homestead. Alan disposed of the *dongo* on the market. The three women denied the new owner permission to occupy the *dongo* and reclaimed it as their arable land. Alan remarried after a year. He had to live with his new wife in his mother's homestead, as the three women would not let him occupy the *dongo*. Alan and his wife later moved to a *dongo* that belonged to a person who had moved to the urban areas. The example highlights how people consolidate homesteads and in the process increase the *matongo*. The case provides an example of how overlapping claims to a *dongo* are managed and also shows that women do not always lose out to men in land-related disputes over land held under customary systems. The case also shows how young men who fail to secure livelihoods resort to *matongo* and also illustrates how some people's management of rural-urban connections creates economic, residence and livelihood opportunities for others.

Chapter 4 showed that some adult married men in Makuku and Ndamba villages who had been allocated land were failing to develop independent homesteads. Mary's case illustrates how one of these men coped by residing with his widowed mother. The men's undeveloped homesteads contributed to *matongo* in Ndamba. There were also such *matongo* in Makuku. Some of the men who had been allocated land but were unable to develop it had disposed of the land on the illegal land market. The focus group discussions and key informants were divided on whether such land was *matongo*.

Alice's case demonstrates how individuals create *matongo* through mobility of people within the village. Each time they claimed virgin land while maintaining the previously occupied place as a *dongo*. Some of the reasons for this migration are failure to strike water when digging a well or poor fertility of the soil, as typified by the Ndamba village head in the Alice case-study. In such cases, people leave their original homestead but maintain control over the *dongo*.

The high mortality rates induce high mobility, which makes it difficult for the authorities to apply the procedures in place. This is because of the fast pace and unsystematic manner at which people move in and out of the village. This situation nurtures *matongo,* which provide an opportunity within which single women can negotiate for land access in the communal areas. The next section discusses the specific manner in which the situation resulted in opportunities and constraints for single women in Makuku and Ndamba villages.

6.4 *Matongo* as a resource for single women

Matongo arise as a result of overlapping land access rights. Overlapping rights occur in customary tenure systems where absent land right holders retain a right to return, but this right does not necessarily exclude other people from the land (Shipton and Goheen, 1992:311). *Matongo* are especially important to women as they form the majority of the communal area residents and have relatively limited access to non-agricultural employment in both rural and urban areas. A study of people in Harare established that 51% of the people who maintained land claims in communal areas had not cultivated the land in the previous two years even though there had been good rains (Potts and Mutambirwa, 1990). Tables 6.1 and 6.2 indicate that some of the *matongo* were neither occupied nor used for agricultural purposes during the study period. Some of the very old vacant plots, especially homesteads, had reverted to bush. The changing rural-urban connections resulted in the periods of vacancy being interrupted by temporary occupation of *matongo* by family members ho needed to use the place for various reasons, including seasonal unemployment, illness and loss of permanent employment. Some *matongo* were only used by the owners for burial of the dead. The sudden increase in mortality induced by HIV/AIDS has heightened the importance of communal areas as burial sites. In some urban based families, women usually returned to the rural areas to cultivate during the rainy season. Others hire labour or allow others to use their land. This is a fundamental aspect of urban-based households' attempt to maximise their economic security by maintaining communal area land rights (Potts and Mutambirwa, 1990:685). In Makuku, some of the people usually hired local labour and provided agricultural inputs. Table 6.1 summarises the different users of *matongo* in Makuku. These included employees, relatives and single women. There were no employees staying on any of *matongo* in Ndamba. This was not because of lack of demand, but because of the strict control that the *sabhuku* had on access into the village.

Single women benefit from the *matongo* as guardians, caretakers and homestead users. In the sub-sample of 22 women, three single women acquired residence in male-headed homesteads belonging to their fathers. These women resided as sub-units of the larger unit because they lacked money and networks with which to secure land and establish independent homesteads. The single women like Ann who were allocated land and developed a homestead in their own right were not residing on a *dongo*. The remaining 18 hearth-holds are presented in the Table 6.3. The 18 hearth-holds were hosted on 16 homesteads because some of the hearth-holds lived together on one homestead.

Table 6.3 reveals that hearth-holds accessed homesteads through marriage, natal claims or the market. There was no market-based access in Ndamba because of the *sabhuku's* opposition to it. All single women who had an independent homestead in Ndamba occupied a *dongo* which belonged to their parents or other relatives before them. The institution of *matongo* enabled these women to overcome their lack of access to shelter, their lack of money to buy or of labour of a man to construct an independent homestead. Table 6.3 also shows that nine of the 18 hearth-holds gained access to residence through marriage. These hearth-holds continued to occupy the homesteads after the death of their spouses. These are also *matongo* according to the criteria in Table 6.2. Ten of the hearth-holds moved to their current place of residence after they had been displaced. The displaced hearth-holds gained residence mainly through natal claims.

Two of the displaced single women gained residence through the market and employment. If all single women-occupied residences are classified as *matongo*, 18 of the 22 single women benefited from *matongo*. If *matongo* only refers to places that were vacant before the single women moved to them, then only 10 of the single women can be classified as deriving benefit from *matongo.* This confirms the importance of *matongo* in providing social security for the women. *Matongo* is a subjective and contested term. The land market increases the number of homesteads through subdivision of existing homesteads. This in turn increases the opportunities for people and single women hearth-holds to negotiate access to residence and community through employment as caretakers and/or temporary use of the homestead.

Table 6.3. Single women hearth-holds' land access through matongo.

Village	Marital status	Residence	Way in which access to *matongo* was negotiated	
			Continued after death of spouse	Moved in after displacement
1. Ndamba	Widow	Marital	✓	
2. Ndamba	Widow	Marital	✓	
3. Ndamba	Widow	Marital	✓	
4. Ndamba	Widow	Marital	✓	
5. Ndamba	Widow	Natal		✓
6. Ndamba	Divorcee	Natal		✓
7. Makuku	Widow	Marital	✓	
8. Makuku	Separated	Natal		✓
9. Makuku/growth point	Widow	Natal		✓
10. Makuku	Widow/Divorcee	Natal		✓
11. Makuku	Divorcee	Natal		✓
12. Makuku	Widow	Marital	✓	
13. Makuku	Widow	Work		✓
14. Makuku[1]	Widow	Marital/natal	✓	✓
15. Makuku	Divorcee	Natal		✓
16. Makuku	Widow	Marital	✓	
17. Makuku	Widow	Marital	✓	
18. Makuku	Widow	Market		✓

[1] Case number 14 refers to a woman whose husband purchased a piece of land in Makuku and lived with her there. When he died, the house they had occupied fell apart. The woman vacated her marital homestead and moved to a *dongo* that had a better house. This explains why the case has both marital and natal access.
Source: Compiled by author from field data.

The women who gained access to *matongo* in Ndamba negotiated their access directly with the *sabhuku*. The *sabhuku* used his position to accumulate land and create patronage relations with a variety of women. The following analysis of Ndamba village *matongo* demonstrates the ways in which *matongo* facilitate the co-existence of multiple claims:

6.4.1 Case 6.2: opportunistic *sabhuku* maintaining control of land

The *sabhuku* in Ndamba controlled six homesteads. The first homestead was the one on which the *sabhuku* and his family resided during the fieldwork in 2005. He was allocated this land by his father who was the *sabhuku* when he married. This is not a *dongo*.

I. The land parcel adjacent to his homestead. The current *sabhuku* allocated this land to his son when the son married. Unfortunately, the son died before he had developed the homestead. The son's widow returned to her natal village with her young son leaving the homestead as a *dongo*. The *sabhuku* reclaimed this land in the process of upward inheritance.

II. *Dongo* two was originally allocated to the *sabhuku*'s sister who had returned from a failed marriage. She since died and the *sabhuku* allowed his other divorced sister (Rita in Chapter 5) to occupy the homestead. The homestead had been inherited by the *sabhuku* as he was the surviving male heir. Rita's claim to this residence was that she was a daughter of Ndamba village.

III. The third homestead is the one involving Alice which the *sabhuku* occupied, vacated, allocated to Alice's husband and subsequently reclaimed on the death of Alice's husband. The *sabhuku* acquired this homestead through his power to repossess as a village head and his claim on it that he originally resided there. Alice's claim was that she was the wife of the deceased male owner. At the end of the research, this homestead was an abandoned *dongo*.

IV. The fourth homestead was a *dongo* previously occupied by the *sabhuku*'s parents who are now deceased. The place was occupied by the *sabhuku's* aunt (Tracy in Chapter 5), a returned, widowed elderly woman. In reference, the place belonged to the '*sabhuku's* father'. This was another single woman-occupied *dongo*.

V. When the *sabhuku*'s mother died, his father married a new wife and moved to a new homestead, as explained earlier in the chapter. In 2005, the *sabhuku*'s stepmother was occupying the homestead as she continued to live there after her husband died. The *sabhuku* controlled this place, as he was now the 'father' of the family. In reference, the place belonged to the *sabhuku*'s stepmother. The *sabhuku* gained control of this homestead through inheritance as the male heir. This too was a *dongo* occupied by a single woman.

The *matongo* transactions described in Ndamba demonstrate the different ways in which single women gained access to *matongo* in Ndamba. The examples show how male and female claims to homesteads co-existed without being mutually exclusive. The cases also illustrate how women's claims relate to the male claims to the same *matongo*. *Matongo* provided immediate relief to single women in sometimes desperate situations like that of Rita, the *sabhuku*'s sister on plot II. The vulnerabilities of single women on *matongo* displayed in relation to *matongo* I, II, III and V are discussed in the next section. In what testifies as an acknowledgement of the increasing demand for shelter by women who returned to their natal homes, the Ndamba *sabhuku* joked about *dongo* IV as the 'women's *dongo*'. Similar attitudes were expressed in Makuku as related in the following example.

6.4.2 Case 6.3: *matongo* as a fallback for single women

A woman born in Makuku became married and moved to her husband's village. She lived in her husband's village for ten years but failed to have children. She returned to her natal village Makuku where she resided with her parents until they died. She continued to live at this homestead and left when she remarried. However, her brothers claimed to be keeping the *dongo* for her in case her marriage broke down again and she returned to the village.

The abandoned *matongo* sites were a form of common property for the villagers. The open access on *matongo* meant that women could collect firewood, wild vegetables, fruit and thatching. This was important, as the village grazing and forestry in Makuku had been converted to private homesteads. The homestead *matongo* were located within the residential part of the village, whereas arable lands were located further away. Access to more cropping land in the residential area enabled women constrained by domestic chores and labour shortages to increase their agricultural production. A single woman had abandoned her arable field and was concentrating her efforts on her own homestead and the adjacent homestead land, which belonged to her deceased parents-in-law. Cathy in Ndamba, who resided in her parents' homestead, used the land on a *dongo* for subsistence farming. This gave her some autonomy.

Ndamba village derived more benefit from the natural resources of *matongo* because their under-utilised land had not been sold on the market. This increased the village's forest and natural grazing resources. The village's *matongo* were a source of materials to construct shelter, firewood and pasture. The availability of these natural resources in close proximity reduced competition and increased opportunities for hearth-holds to access firewood, thatch and grazing.

The increase in the crime rate necessitated caretaking arrangements in Makuku village, as property that was abandoned or left unoccupied was vulnerable to looting. Caretaking arrangements ranged from the specific employment of an individual or a family member to reside on the place to having somebody, usually a relative, in the local community keep an eye on the homestead. Single women in Makuku occupied *matongo* as guardians of the homesteads for absentee owners. The old man cited in Section 4.6.2 is an example of how men also benefit from *matongo*. Women also benefited as caretakers of *matongo*. The case of Timika in Makuku demonstrates how women can gain from such arrangements.

6.4.3 Case 6.4: living off *matongo*

Timika was a widow who lived with her mother in Makuku. Timika was born in Makuku. She got married in the adjacent village and moved from Makuku to the neighbouring village. The husband's parents were deceased. Timika and her husband used the husband's deceased parents' homestead *dongo* for residence and farming. Timika and her husband divorced. Timika returned to her mother's residence in Makuku. She then left for Harare to work as a housemaid. Timika stopped working to marry another man. They lived on the commercial farm where her husband was working. Timika's husband died and she returned to Makuku where she resided with her mother. An immigrant bought a homestead, developed it and employed Timika to live on his *dongo* while he worked in town. Timika's employer provided her with agricultural inputs. The vulnerability of single women hearth-hold's *matongo* resource access, and specifically that of Timika, is explored in Section 6.6.1. However, this excerpt illustrates how Timika's livelihood straddled the rural urban space as a single entity within which she moved to earn financial income and fall back when the marriage and employment failed respectively.

In another example, a single mother of three children was residing on a *dongo* in Makuku. The woman had been employed to look after the homestead by the owners who lived in Harare. This woman was paid a wage, which she combined with her trading income to look after her children. The woman explained that her predecessor had moved into a place of his own in Makuku. She was saving money so that she could purchase land to construct her own homestead.

The opportunity to take care of a *dongo* increased opportunities through which people could gain access to homesteads in the village. Makuku had male and female caretakers whose had since married and were raising their families on *matongo* under their charge.

6.4.4 Case 6.5: *matongo* as a source of shelter

Tessa was a 52-year-old widow who resided in Makuku. She was the *sabhuku*'s sister, and lived with her eighteen year old son and a five year old nephew. The house was a well-constructed but ill-maintained brick structure. There was a well, which had collapsed. Tessa grew up in Makuku village. She married a commercial farm worker. Tessa and her husband negotiated with her brother, who was the *sabhuku* in Makuku, for space to construct their residence. They constructed two huts. They had three children. Tessa's husband fell ill and died. Tessa's huts started to disintegrate and she could not maintain them. Tessa negotiated with the *sabhuku* for permission to occupy a *dongo* that belonged to a now deceased sibling of Tessa and the *sabhuku*. The deceased man had constructed a solid brick and mortar house. His wife and young children left the village after his death, leaving their house a *dongo*. Tessa and her sons moved into the 'new' house and had better shelter. Tessa maintained her own *dongo* and used it for agriculture. Tessa's sons were reluctant to invest in either of the two *matongo*. They feared the risk of eviction from their mother's relatives.

These cases illustrate how single women negotiated access to shelter and residence through *matongo*, and thereby reducing the vulnerability highlighted in the previous chapter. However, this does not mean that all single women had access to the land they needed or that the land was equitably distributed. The co-existence of claims facilitated complementary access to land and residence for different claimants.

The village of Makuku was experiencing land shortage because of the high land demand. Table 6.1 reveals that some people used *matongo* for cultivation. Relatives used *matongo* for free, whereas non-related people entered into some kind of lease. Single women in Makuku also accumulated land through *matongo*. One woman gained access to two *matongo* through upward inheritance of lands belonging to her deceased sons. There was an elderly woman in Makuku who had acquired control of several homesteads through *matongo*.

6.4.5 Case 6.6: Ann: land baroness on *matongo*

She had:

- control over her own plot which she was allocated in her own right by her father who was the *sabhuku* when she returned to the village after the end of her marriage. This was not a *dongo;*
- control of the *dongo* which belonged to her son who moved to the former commercial farms after he was allocated a piece of land following the fast track land redistribution exercise;
- control of a *dongo* which belonged to her deceased brother;
- control of her deceased parents' *dongo*.

The ways in which the woman gained access to these *matongo* show the potential influence of rank, as she seemed to have inherited land in the same ways as, and in some cases ahead of, the men like the *sabhuku* who was her nephew. Ann controlled the largest land holding in Makuku village. Although she was too old to utilise this land fully, she benefited from the land as she collected firewood and thatch, and she leased the arable land to some landless immigrants in exchange for some of their harvest.

There were widows who remained in Makuku after the death of their husbands and continued to use the homesteads and arable land. Some of these women sold[61] the arable land and/or subdivided their homesteads and sold the remainder to immigrants. This enabled women to derive another benefit from *matongo*. The evidence of widows disposing of 'inherited land' on the informal market proves that they have more decision-making power than could be claimed on the basis of normative customary rights.

6.5 Vulnerability of women's land access through *matongo*

Access to a *dongo* is based on fragile claims which rest on underlying stronger land claims. This is a source of both the weakness and the strength of *matongo*. The main threats to *matongo* access originate from changes in the underlying relationships upon which the *matongo* rest. These changes are contingent upon other changes introduced by the State through policy and legislation, the market through supply and demand dynamics, and mortality and people's mobility patterns in response to these changes.

In Makuku, the deteriorating relations, due to the increased competition over land and grazing, between the people who had purchased land on the market and those who viewed themselves as original inhabitants were a source of instability. One of the consequences of this in customary tenure areas is a process of narrowing the definition of belonging (Peters, 2004:301). The people who see themselves as belonging use metaphors of 'stranger' and 'squatter' to refer to those denied legitimate status by the landholding community (Nyambara, 2001a; Peters, 2004). In Makuku during the heated debates and meetings, single women were sometimes referred to as squatters by local squatter resistance commitee. The men argued that these single women did not belong to Makuku because they should have worked at making their marriages a success and then would have remained in the marital villages. The women who gained access to residence in Makuku on their natal claims were constantly threatened by these young men, and their names had been included on the list of those to be evicted along with the people who purchased land. The autochthons in Makuku also pressured that *matongo* land be reallocated, especially to their sons. The locals resented *matongo*, which they viewed as providing an opportunity for migrants to enter and eventually settle in their village. The autochthons gave examples of at least three former caretakers who had now married and/or acquired their own place in the village in the previous three years. The treatment of single women and immigrant men suggested that these two categories occupied similar positions in the land access hierarchy.

[61] The fact that the community let these women dispose of the land rights on the market means that they were recognised as having some power to dispose of the land.

The under-utilisation of arable land caused by the lack of access to agricultural inputs rendered the land vulnerable for reallocation or sale by the powerful, like the Makuku *sabhuku*. The Makuku *sabhuku* tried to sell some of the land belonging to Ann, the land baroness, his land-rich aunt, but she resisted. The woman explained in her own words how she resisted attempts by the *sabhuku* to sell some of the land under her control;

> *'I pulled up the pegs from my field and threw them away.'*

Chapter 4 described how the Makuku *sabhuku* allocated his migrated brother's *dongo* to the village gardening co-operative. The initiative was stalled after the *sabhuku*'s widowed mother refused to allow the co-operative to occupy the land. The practise of *sabhuku* allocating fallow land to other villagers was also observed in Buhera (Andersson, 1999:554). The Ndamba women's *matongo* access was not under similar threat because of the *sabhuku*'s attitude towards the market. The Ndamba *sabhuku* viewed fallow land as a common resource providing firewood and pasture for the population. However, the Ndamba *sabhuku's* control of transactions in the land means that single women hearth-holds in Ndamba cannot benefit from the land market through for example disposing the 'excess' land or subletting to land short people.

High mortality has increased *matongo*. Some women accessed *matongo* belonging to people who have moved out of the villages in response to the various stimuli. These women' access is vulnerable to the death of the owner of the *dongo*. The women would be displaced by family members and/or local leadership seeking to take advantage of the opportunities created by the demise of the owner as demonstrated in Alice's experience (see Section 5.2.1). Chapter 5 described the existence of nested hearth-holds consisting of more than one single woman residing on a homestead: cases 7-9 and 15-16 in Table 6.3. In these cases, daughters were living on homesteads headed by their widowed mothers. In such cases, the mother had less tenous land rights than the daughter and her children. These daughter hearth-holds were vulnerable to the death of the senior woman, as this would destabilise their claims.

The men who accessed *matongo* land were equally vulnerable. Employees were vulnerable in the same way as women caretakers. Some male children in Makuku resided on *matongo* with their mother. The mother had accessed *matongo* on a natal claim. These men were vulnerable to eviction on the death of the mother. This is the reason Tessa's sons in case 6.5 above were reluctant to invest in the *dongo* that their mother occupied. The men coped by purchasing their own homesteads and not investing in *matongo*.

Single women's claims to *matongo* remained vulnerable to male claims on the same land. The description of Ndamba highlights the juxtaposition of the *matongo* interests of the Ndamba *sabhuku* and those of single women in his village. In all five cases represented by homesteads II to VI (Case 6.2 in Section 6.4.1), the *sabhuku* had a male claim to the *dongo* which was stronger than that of the respective single women because of the *sabhuku's* capacity as a former owner, male heir and village head. The case shows that, in instances where male and female interests are in conflict, male interests dominate. This led to the eviction of single women as typified by Alice in homestead III and the village head's son's wife in homestead I in Case 6.2. Experiences in Makuku on the other hand show that women's interests can supersede those of the males. Examples of this are Ann (Case 6.6) who resisted the *sabhuku*'s attempts to dispossess her of *matongo* in her control

People who had escaped the war by migrating to urban areas returned after the cessation of hostilities to reclaim their homesteads and arable fields in Gokwe (Nyambara, 2001a:258). The expense of urban life, coupled with the added burden of HIV/AIDS, resulted in the re-activation of rural communal land claims by those whose urban livelihoods were threatened. Research carried out in other communal areas of Zimbabwe concluded that urban migrants returned to their rural homes after structural adjustment programmes. People who moved out of the villages to settle on former commercial farms after the land resettlement programmes also returned to the communal areas if they failed to meet their basic needs in the resettlement area (Bird and Shepherd, 2003:596; Potts, 2000a:814, Chimhowu, 2002:568; Chimhowu and Woodhouse, 2006a:738). Operation Restore Order displaced people from the urban areas. During the fieldwork, five victims of operation Restore Order returned to the villages. Although not observed in this study, the return of immigrants to communal areas may have led to the displacement of people occupying *matongo*.

6.6 *Masabhuku*: defining, managing and disposing of *matongo*

The uneven enforcement of legal codes and ambiguous customary paradigms that govern communal area resources create gaps and opportunities for people to manipulate the system (Nyambara, 2001a:257). The time of the research, Mwanza ward was experiencing high land demand as a result of changing rural-urban connections and the displacement of people by the fast track land reform programme and operation Restore Order. As land became more contested, people devised strategies and took advantage of loopholes in the system to increase their control over land. It is usually the powerful and well-established people in the village, such as the *sabhuku*, elderly and powerful individuals who can do this (Shipton and Goheen, 1992:311; Nyambara, 2001a:261). The *masabhuku* in Ndamba and Makuku took advantage of the vague process leading to delineation of land as *dongo* to accumulate land and redistribute it as part of their patronage system. The manner in which the Ndamba village head used *matongo* to accumulate land is demonstrated in Case 6.2. A *sabhuku* in Gokwe converted the *zunde* (see note 3) field to personal property[62], which he used to grow cotton (Nyambara, 2001a:262). The Makuku *sabhuku* used his position to establish Makuku 2 and Makuku 3 by selling land to immigrants. The following description of Makuku *sabhuku*'s residential history reveals how the *sabhuku*'s actions created risks and opportunities for the single women and immigrants in the village.

6.6.1 Case 6.7: *sabhuku* taking advantage of *matongo*

The *sabhuku* inherited the position from his late father. By the time his father died, the *sabhuku* was a married man who had been allocated his own homestead in Makuku. When he assumed the *sabhuku* position he had power to allocate land. The *sabhuku* sold his homestead to Chako, an immigrant, and moved onto an unclaimed place where he set up home. The purpose of briefly living there was to formalise the claim before selling it to another immigrant. Meanwhile Chako lived in town, but he developed the homestead that he purchased from the *sabhuku* with the hope of retiring to it at the end of his working life. Chako hired Timika, a divorced adult woman who returned to the natal village after her divorce, to look after the homestead. He provided her with agricultural inputs. The arrangement lasted for eight years.

The *sabhuku* then moved in with his sister (Tessa). The sister was living on a *dongo* which was left vacant after the death of the *sabhuku*'s brother. At the time of his death, the deceased brother had divorced his wife and she went away with her two sons from the marriage. Tessa claimed to be keeping the *dongo* for those sons. Tessa moved into the homestead to take advantage of the shelter when her own shelter deteriorated. The *sabhuku* moved out and claimed another piece of virgin land, which he briefly occupied and then sold. By then the village had run out of new land. The *sabhuku* carved a place for himself on the water channel (land reserve set aside for the management of rainwater drainage). This was all in a space of ten years.

The *sabhuku* began to feel the pressure of living on an illegal plot and, having sold all the land at his disposal, devised a strategy to reclaim the land from Chako. The *sabhuku* claimed that, since Chako was not living on the land, he as the village head was repossessing it. By the same action, the *sabhuku* claimed to be repossessing the *dongo* that he had previously occupied. The *sabhuku* ordered Chako's caretaker, Timika, to vacate the premises. When Chako learnt of this development, he demanded that the *sabhuku* refund all the money that he paid for the purchase and improvements on the homestead. In a public gathering, the *sabhuku* denied selling the land to Chako. The *sabhuku* asked the villagers to endorse his actions but the villagers did not do so. Nonetheless, although they generally felt that the *sabhuku* was acting

---

[62] The *zunde* programme was undermined by the colonial government through systematic destruction of the traditional leadership's authority and the assumption of the food security guardian role by the State. When the *zunde* concept failed, the *zunde* land remained unused and became a *dongo*.

unfairly, they were scared that he would abuse his powers as village head to victimise them, e.g. deny them access to food aid. Alledgedly, while this was going on, the *sabhuku* had sold the homestead which he had been occupying. Chako reported the story to the sub-chief. In the meantime, the *sabhuku* took up occupancy of Chako's homestead. The sub-chief ruled in favour of Chako on the basis that the land had been continuously occupied and that there were a lot of unutilised *matongo*, which had been standing for longer, which the *sabhuku* could repossess. The *sabhuku* vowed to remain on the homestead. The case was referred to the Chief for arbitration. The story had not been concluded at the time of finishing the fieldwork, but Timika was still residing on her widowed mother's homestead.

This case illustrates the complicated processes underpinning customary land access in combination with vernacular land markets. It demonstrates the tenuous rights that women and immigrants have through *matongo*. *Matongo* provided livelihood diversification opportunities for the Makuku *sabhuku*. The case shows male immigrants' land rights to be as tenuous as those of single women in the communal areas. The migrant's case to defend his claim to the land was weakened by his failure to use the place continuously, lack of proof of a financial transaction, the illegality of the transaction and the failure by the Makuku community to support his land claim.

The behaviour of the *sabhuku* highlights the arbitrary character of the legal environment of the communal area. This created uncertainty and opportunities for abuse of power by the *sabhuku* and increased the vulnerability of his sister, other single women and possibly all people occupying *matongo* in Makuku. The *sabhuku*'s sister Tessa could potentially lose her access to the homestead if the *sabhuku* decided to claim his male inheritance right to it. Even the *sabhuku*'s wife was potentially vulnerable as both she and her husband had failed to establish a clear claim to any homestead in the village. As a result, she could potentially lose her homestead if her husband died before her.

The cases illustrate the fact that, if there is a contestation between males and females, the risk is that female rights will be subordinated. As a result, single women generally avoid contestations. The cases expose how people in authority may use their power to their own advantage and dispossess those with less power. The *masabhuku* used their power and community control to dispossess single women. The evicted women resorted to other *matongo*.

The absence of a land market in Ndamba meant that there were fewer homesteads and consequently fewer, and less diverse, opportunities available for *matongo*. The Ndamba *sabhuku*'s reaction to reduced agricultural production was to convert the land into a community forest and paddocks. This was a different decision to that of the Makuku *sabhuku* who decided to make money out of the resource. The result was that access to fruit trees, land and grazing on unused *matongo* was better for the single women in Ndamba. *Matongo* provided firewood, grazing and forestry products. This was important in mediating women's vulnerability as they did not have to take the same risk as people in Makuku who, because their forest was depleted, resorted to stealing from the commercial farms. The centralised governance of land, homesteads and *matongo* in Ndamba resulted in more stringent rules governing the access of single women to land. The stricter application of rules also resulted in the reduction of the number of *matongo* available in Ndamba. The land pressure in Makuku increased conflict and tension between the various groups of people and resulted in the targeting of single women as squatters. Ndamba village did not have this problem because it did not have such a large population.

There is a lack of clear legislation from the Zimbabwe government to guide authorities in their governance of *matongo* and land in communal farming areas in general. According to the Native Land Husbandry Act, it is legal for the local leadership to repossess and re-allocate land belonging to absentee owners. This has led to a loss of communal land rights through non-use (Machingaidze, 1991:578). This provision has not been followed consistently in the communal areas of Zimbabwe. Berkvens (1997:3) observed that people could lose their rights to land when they did not use the land. Pankhurst's (1991:612) study in a village in Mashonaland East concluded that rights to communal land could not be lost through non-use. The inconsistent application of the law has created uncertainty and maintained the vulnerability of *matongo* land access. The outdated Native Land Husbandry Act has not been replaced by updated legislation.

The urban migrants interviewed on the security of their land rights in communal areas were divided, with some feeling that their land rights were not threatened, whereas others felt that the land could be repossessed. People who were living in urban areas but maintained communal areas land rights feared that they could lose *matongo* to reallocation by the *sabhuku*, whereas others feared that somebody would just start using their land without permission. Both groups, however, linked this not to government policy but to the land shortage problem (Potts and Mutambirwa, 1990). The Mwanza ward *masabhuku* concurred in a focus group that they were empowered to repossess and re-allocate land on the basis of non-use, i.e. after it had been a *dongo* for more than three years. The ward vacant land register, however, showed that, in all the villages, there were *matongo* that were more than three years old, over which the *masabhuku* had not exercised their repossession and re-allocation powers. Table 6.1 shows that only five of the *matongo* were less than three years old. The rest had been *matongo* for up to 20 years, but the *masabhuku* had not repossessed them. The *masabhuku* admitted that repossessing and/or reallocating *matongo* was not an easy task because of the kinship networks that existed between a *sabhuku* and some people, fear of witchcraft and related practices by the *sabhuku*, especially when the place concerned became a *dongo* because of death. The *dongo* of a person who had moved out of the village was never repossessed as long as that person or his descendents were believed to be alive and could possibly return. This suggests the existence of checks and balances in the system that protects land claims from arbitrary possession by the *sabhuku*.

The improvements that people made on their land while it was *dongo* made it difficult for the *masabhuku* to repossess the homesteads, because whoever took over the land would be obliged to compensate the owner for his/her investments. This is only possible when the owner can be located and found willing to dispose of the place. The *sabhuku* who displaced Alice got round this problem by asking the deceased man's father to remove all his late son's improvements from the *dongo* before he reclaimed it. Even in instances where there were no improvements, a *dongo* owner's relatives who lived in the village fiercely resisted any attempts by the *sabhuku* to repossess and or reallocate *matongo*. The relatives claimed to be acting on behalf of the rightful owners of the land.

The actions of the senior women who resisted the *sabhuku*'s attempts to sell their *matongo* exhibited the agency of single women in defending their land claims on *matongo*. Women also resisted through collective action, as they did when they resisted Alan's attempts to sell the land (Case 6.1, Section 6.3.3). Alice and Timika's decisions after their eviction from *matongo* show another strategy that women use to cope with dispossession. They avoided confrontation with the *sabhuku* who threatened their access and negotiated access to other *matongo* in the village. The constant request by single women for residence access through *matongo* has resulted in a change in the gendered resource allocation in the communal areas by the setting aside and maintenance of homesteads by the kin groups for use by the single women and their children. However, the status of children remains a sticking point.

6.7 Conclusion: bargaining for land within patrimonial governance systems

The focus on practices and individual single women hearth-holds rather than norms highlights an increasingly important mechanism through which people gain and maintain access to land and residence in the communal areas of Zimbabwe. Norms marginalise *matongo* and treat them as a temporary phenomenon; yet observed practice illustrates that *matongo* are widespread and their existence is an enduring phenomenon. The different ways of delineating *matongo*, the diverse experiences of men, women and *masabhuku*, and the gaps between norms and practices illustrate the ways in which general things are different in ways that matter to the individuals concerned.

Matongo increased the ways in which single women marginalised in other land access modes gained communal area resource access in the study area. Both men and women benefited, but the resource was more important for women because of their historical disadvantage and gendered governance of land, community and natural resources in a customary tenure area. The diverse rural-urban connections, mortality and market activity in communal areas have continued to promote the existence of *matongo*. The practise by people leaving communal areas of retaining their communal land claims and the 'accumulative' tendencies of those remaining in the village highlight an effective strategy used to cope with uncertainty in a fast-changing environment. The people who were in a position to do this were elderly men and women in the village. Senior single women increased their control of land and homesteads through *matongo*. The women used their control as a basis for negotiating access to employment and community membership in the communal areas. Through control of *matongo*, some senior women were able to dispose of or lease out land to land-short people, especially those who had bought land on the market. In return, the women gained access to financial capital, labour and agricultural produce.

The emergence of the informal land market in Makuku may have served to reduce the availability of *matongo*. The experiences of the Makuku community show that the phenomenon of *matongo* persists despite increases in mortality rates, increasing mobility of people moving between urban and rural areas and the increased importance of market transactions, because of the plurality of processes that give rise to the phenomenon. The informal land market has increased the diversity of ways through which women can benefit from *matongo*, including caretaking or employment as a means of gaining access into the community. Such opportunities did not exist in Ndamba. The prohibition of the land market in Ndamba ensured the availability of kinship-transacted *matongo*. This was a valuable resource for single women who were displaced. The decision by the Ndamba *sabhuku* to turn *matongo* into forest and grazing benefited single women, as they were more reliant on the natural capital.

Matongo land access in communal areas of Zimbabwe is fraught with contradictions: the *matongo* layer of land access has emerged out of the legal uncertainty, high mobility, mortality and rural-urban connections. These processes have maintained the *matongo* layer and provide invaluable opportunities for the landless and marginalised people in communal areas. The same processes of legal uncertainty, rural-urban connections, high mobility and mortality are also the source of *matongo* fragility. The fragility of the *matongo* made women vulnerable to opportunistic behaviour by *masabhuku* in the two study villages. The two cases of the *masabhuku* illustrate that they used counter claims on *matongo* to destabilise the claims of single women. The *matongo* concept is also paradoxical: on one hand, *matongo* land access reinforces the patriarchal structure of customary land access which maintains male superiority. On the other hand, the ability of women to defend *matongo* claims against male kin shows that kinship claims of women to *matongo* are stronger than commonly believed. The rural urban mobility caused by HIV/AIDS, structural adjustment, fast track land reform and operation Restore Order have led to the return of migrated men and a reduction in the out-migration of men, and have resulted in the loss of *matongo* access by all those who depend on them to access land and gain village membership. However, the high HIV/AIDS-induced death rate for the moment acts as a leveller. The fact that people who are expelled from *matongo* generally find another *dongo* to occupy shows that, for now, *matongo* opportunities exceed demand. This means that this opportunity will continue to be available and act as a pressure valve in the immediate future. Single women's *matongo* land access is not equally vulnerable. Elderly single women have stronger claims to *matongo* than junior women and men. In addition to single women, other categories of women and men are vulnerable in communal areas: immigrant men and the children of the single women who acquire resource access through *matongo*.

The multi-layered nature of rules governing land access in communal areas is a valuable resource for women, given the highly uncertain conditions under which they have to make decisions. *Matongo* is a mechanism for ensuring a more efficient land distribution system as it helps to address inefficiencies inherent in the customary and market-based land distribution systems. The diverse situations that give rise to and perpetuate *matongo* maintain this resource even as conditions change.

This examination of *matongo* has enabled the analysis of land claims as complementary rather than mutually exclusive competing categories. The terms for gaining access to *matongo* are not always clear. In some cases, the conditions are quite restrictive by, for example, denying women autonomy to decide with whom to co-reside. However, *matongo* provide a valuable resource and relief for single women who are faced with homelessness and impoverishment when they are displaced. Single women's claims to *matongo* are stronger than presented in normative discussions on single women's resource access in communal areas. The *matongo* discourse shows that the male lineage has a kind of obligation to accommodate their single women. This is the main reason why returned women are relatively more secure than women occupying *matongo* through marriage, employment or as a guardian. *Matongo* act as a pressure valve by giving access to the desperate people, while retaining the claims of the absent people.

The increased mobility and mortality has significantly changed the way in which women gain access to homesteads. The extent of a single woman's security of tenure on *matongo* depends on her health, the way in which she gains access to *matongo*, her rank and the decisions she makes. *Matongo* access is destabilised by death, rural-urban connections or illness. The indeterminate rules governing *matongo* are a source of uncertainty for people who gain access to land through *matongo*. This is a source of vulnerability. The unclear governance of *matongo* results in some people retaining ownership, while others dispose of their land claims on out-migration. Table 6.1 shows that purchased properties are also a source of *matongo*; this means that this phenomenon transcends market, customary and employment transactions of resources. The existence of *matongo* shows that single women are not the only people who have tenuous claims to land. The claims of people who purchase land in the village and children of both sexes belonging to single women are shown to be weaker than those of single women hearth-holds on *matongo*.

The specific local context is a significant determinant of the numbers, processes and vulnerability of *matongo*. The Makuku example shows how the land market increased the diversity of ways by which people accessed *matongo* in their local area, but it also increased conflict between the different people. The Ndamba style of governance shows how *matongo* can be managed to provide increased social security not only for single women on the homesteads, but through promoting access to fruit, thatch, building materials and grazing on the fallowed land. However, this limits women's access to residential land – an access that is important for negotiating access to community and residence.

Matongo is an increasingly important way through which single women negotiate residence, land and community access in the gendered resource governance space of the communal areas. *Matongo* enable women to overcome the barriers imposed by the market and custom, and to gain access to resources in the communal areas without directly confronting the patriarchal norms.

Chapter 7

Livelihood decision-making experiences of single women-directed hearth-holds[63]

7.1 Introduction

Communal areas serve as residence and livelihood bases for single women. Chapters 5 and 6 have focused on how single women gain access to residence and community in communal areas Since hearth-hold residential access is the basis for negotiating access to financial capital, human capital and access to other livelihood resources it is important to understand how single women make livelihood decisions. Through a series of livelihood decision-making histories, this chapter shows how single women piece together different kinds of livelihood capitals to confront (not always successfully) the insecurities of land, health, the economy and livelihood in rural Zimbabwe. The analysis draws from life histories of 22 hearth-holds who participated in the in-depth sample. Life histories of single women hearth-holds provide the longitudinal data through which influences of macro-level policy changes can be judged. The historical analysis of the pathways unravels the single women-hearth-holds' experiences. The rural-urban connections was used to understand how the complimentary relationship between rural and urban areas evolved and influenced single women's livelihood decision-making This is triangulated with data collected through focus group discussions. The exploration of the day-to-day activities of people has enabled the identification of the range of factors that impinge on decision making in a more realistic way (Scoones *et al.*, 1996:73).

7.2 Contextualising hearth-holds' livelihood decision-making environments

Hearth-holds construct livelihoods in conditions characterised by exposure to multiple shocks. The concepts of livelihood vulnerability and pathways will be used to analyse the life stories of the hearth-holds. The concept of vulnerability is useful for unravelling how hearth-olds experience threats and how they manage the threats to their livelihoods. Vulnerability can also be used as forward looking concept that describes how prone individuals are to being unable to cope with certain adverse events that may occur. The environment in which hearth-holds construct livelihoods has been shaped through a long history of transformation and rapid changes over

[63] A version of this chapter is has been submitted for publication under the title 'Single Women's Experiences of Livelihood Conditions, HIV and AIDS in the Rural Areas of Zimbabwe' in: Niehof, A; G. Rugalema and S. Gillepsie (eds). *AIDS and Rural Livelihoods: Diversity and Dynamics in sub-Saharan Africa*; Earthscan, London (forthcoming).

time. The concept of pathways which reflects how individuals cope with vulnerability in dynamic conditions will be used to study how hearth-holds make decisions over their life course. The units of analysis will be the hearth-holds, homestead and the village. Hearth-holds' livelihood portfolios are composed of activities that are pursued in the communal and urban areas. In order to take into account the spatial spread of hearth-hold livelihood activities into areas beyond communal areas, the concept of rural-urban connections is utilised. Connections refer to the diverse movement of people, goods and services on the rural-urban continuum. The connections acknowledge that the rural and urban areas exist as a single social universe.

The communal area residents' capacity to pursue various livelihood activities is undermined by the scarcity of financial capital in the rural area. The liquidity of communal areas declined when the urban economy contracted and wage income, which had formed the core of urban rural monetary flows, declined. Old age pensions and rental income, previously a secure source of income, have been undermined by hyperinflation. Women's reproductive responsibilities and the increase in morbidity rates have limited women's capacity to diversify income-generating activities. Single women are marginalised in their access to non-agricultural income and labour, seeds, and draft for agricultural production. The State limits communal areas inhabitants' access to non-agricultural income through strict controls on activities and the historical bias of education and mobility in favour of men. The availability of stocks for trading is uncertain due to the erratic supplies, which characterise the Zimbabwean consumer market. The inflationary environment and high transport costs have had a negative impact on the returns from trading activities. The sale of home-brewed beer remains illegal under Zimbabwean law. Beer brewers risk having their beer confiscated by the law enforcement agents.[64] Pottery, a diversification option for elderly women, is a seasonal activity and remains vulnerable to the limited viability of cash in rural areas. The government, which aims to provide communal area households with food and agricultural inputs, has been struggling to cope with demand.

The increasingly difficult environment in the communal areas has fuelled beliefs about misfortune and witchcraft. This, in turn, has led to increased reliance on spirits, and consequently an increase in the incidence of traditional beer brewing. The beer-brewing activities are strictly regulated by customs. Postmenopausal women are specifically central to this ritualistic beer brewing and earn a livelihood from the activity. The Christian churches also provide financial, moral and material support to their members (Scoones *et al.*, 1996; Vijfhuizen, 2002; Hartnack, 2005). Some churches have a reputation for healing. They usually provide free assistance. The high morbidity has increased church membership as people seek healing (Vijfhuizen,

[64] At the time of the research, the beer brewers feared that Zimbabwe Republic Police members would dispose of any beer brewed and sold in the communal areas.

2002:221). Conversely, high mortality has decimated church membership. This in turn undermines the church's capacity to extend moral and financial support to the remaining members. The impacts of the shocks and constraints on the different households are mediated by the individual circumstances. The single women in the study took (sometimes) risky decisions in an effort to maximise opportunity and safeguard their hearth-holds.

7.3 Hearth-holds' livelihood activities

7.3.1 Reproductive activities

Women are responsible for the material reproduction of labour. This includes preparing and producing food, maintaining the home, and bearing and raising children (Adams, 1991a; Pankhurst, 1991; Scoones *et al.*, 1996:48; Huisman, 2005). Traditionally, women's childcare obligations have been towards young children. HIV/AIDS, economic stagnation and livelihood changes have altered the nature of single women's maternal responsibilities in the study villages. Chapter 4 showed that nearly two thirds of the homesteads in the two villages were fostering minor children. The parents of the fostered children were deceased or had moved out of the communal areas to work and/or remarry. The foster parents included a high number of grandparents. Childcare responsibilities included primary care, schooling, and clothing and health fees when necessary. The declining government support has reduced institutional healthcare. This has increased the work burden of women, who are the traditional carers. This has increased the vulnerability of the hearth-hold's livelihood as the responsibilities of caring for the ill have limited women's mobility and opportunities for earning an income and diversification of their activities. Vulnerability in this instance refers to how prone individuals and families are to being unable to cope with uncertain adverse events that may happen to them.

In Chapter one it was discussed that one of the reason people had children was to have them look after their parents in their old age. The declining economic conditions in Zimbabwe negated this trajectory. Male and female adult children have failed to leave the parental home, delayed departure or returned to the homes in which they were born and raised, because such 'children' failed to establish independent homestead, experienced marital dissolution or succumbed to illness. Twelve of the 22 single women in the in-depth sample had dependent adults[65] in their care. These included single and married sons, their spouses and children, and in some cases single daughters and their children. Though hearth-holds are responsible for both young and adult children, the table focuses on the adult children as they have largely been under reported in research that focuses on women's burden of care.

---

[65] Unemployed without income. Though some of them were sometimes erratically employed, the parent/ host was the main provider in the homestead.

Seven of the single women in the sample were hosting their sons and daughters who had returned to the parental home after the death of their spouse, and children who had lost one or both parents. Five of the single women hearth-holds in the sample lived as dependents in other people's homesteads. Three of these lived on the homesteads of other single women. Table 7.1 summarises the obligations of the single women to the adult children.

Table 7.1. Responsibilities of single women to adult children.

Mother's status	Child's problem
Elderly widow	Ill married daughter (28 years) Unemployed, Ill adult son (38 years old)
Divorcee	Adult son unmarried (26 years old)
Elderly widow	Divorced daughter (36 years old)
	An employed son who works and lives at his place of work. The man has not developed his own homestead (34 years old)
Elderly widow	Widowed son (32 years old) Widowed and divorced daughter (26 years old)
Elderly widow	Divorced displaced daughter
	Divorced son who moved back to mother's homestead
Elderly widow	Adult unmarried son
Elderly widow	Unemployed married son who has not developed his own homestead

Source: Fieldwork 2005.

The table shows the circumstances under which adult children remained in or moved back into their single mother's residence. The children's reasons for moving back were marital breakdown, illness and unemployment. Both male and female children returned to the mother's homestead. Divorced and widowed sons moved back to the mother's homestead as they lacked capacity to manage a homestead on their own. The ages of the children show that these are normally the economically most productive group.

Status of child's own household	Duration and timing of assistance
Intact but needs assistance when ill	Whenever the daughter fell ill
The informal business he worked in was destroyed by 'operation Restore Order' and The man subsequently became ill and moved in with his mother	Ended when he died
The informal housing he occupied was destroyed by the government during 'operation Restore Order'	Till he secures new employment
Dissolved	Not possible to say as the daughter might remarry and/or remain single for the rest of her life
Lives in work-tied housing with his wife and children. When he retires and or loses the job, he will have to fall back on his mother's homestead	Fallback/maintenance of rural land rights Life term
Dissolved	Till remarriage of the son
The woman lost her homestead after divorce. She returned to the village She left her children in the village but she resides at the growth point where she works	Not possible to say as the daughter might remarry and/or remain single for the rest of her life
Dissolved	Not possible to say as the daughter might remarry and/or remain single for the rest of her life
Dissolved	Not possible to say as the daughter might remarry and/or remain single for the rest of her life
Delayed departure	Till married and/or employed Unpredictable
Intact: residing in urban immovable property of the mother	Until the son finds employment where he earns enough to develop his own homestead

The table reveals that the mother's responsibility for her children is no longer just a factor of time but also depends on the children's health and ability to gain and maintain access to labour, land, building materials and financial capital to set up an independent homestead, such as marriage and/or employment. The children remained in the mother's homestead for indeterminate periods. This is because with the information at hand it was not possible for the researcher, the children or their single mother to predict if the children would remarry and/or secure employment to enable them to set up and independent homestead in future. The majority of children who returned to their mother's care also brought a dependent grandchild. Single women had to nurse their ill children and provide for them following the break-up of their own household. The ill adult children increased a single woman's burden of care and constrained her opportunities to pursue other livelihood activities. The table also shows that single women hearth-holds had to support children who were employed, because the employed children's wages were inadequate for them to set up independent homesteads. Some of the resident adult children assisted the single women with chores, and sometimes provided money from the opportunities they secured.

However, in all the hearth-holds in the case-studies, the single women were the main providers and decision makers. The adult children assisted their mothers in the homestead with labour and when possible some income from opportunistic activities like trade and participating in government public works programme. The ill children could not assist in the agricultural activities. The adult male children did not participate in agricultural activities preferring instead to focus their attention on securing access to financial capital. The men viewed farming as women's work. The adult male children who were more educated looked down on communal area agricultural work.[66] An example is Ann, one of the women hearth-holds was staying with a divorced son and his daughter. The son, a former liberation struggle veteran was getting a pension from the Government of Zimbabwe. He lived with his mother but spend all his days drinking beer and telling war stories. The mother implored the researcher to report her son to the government as he did not even invest his pension in his daughter's education. He would not help his mother in the field. He sometimes assisted her to repair the infrastructure around the homestead. This man and other hearth-holds' unemployed sons hung around their mothers' homesteads and engaged in occasional casual jobs and if they had livestock, help with the ploughing. They left the weeding, harvesting and planting to the women. These findings corroborate similar conclusions reached by Bryceson and Berkvens in rural Zimbabwe. In a study in rural Zimbabwe, Bryceson found that hardly any of the male or female interviewees stated that they were farmers by occupation

[66] During the homestead visits, I would come across adult children lounging in the shade while their mothers and wives were working in the fields. When I asked the mothers about this, they told me that these 'educated' children did not like to work in the fields.

because 'communal farming is not regarded as a job. It is valued as a residual kind of work ... hard work with low and insecure returns' (Berkvens 1997:26-7). Men were more likely to state that they were unemployed instead of admitting to being a farmer, whereas older farming respondents classified themselves as retirees. (Bryceson 2002: 22).

Chapter 5 highlighted how, sometimes, single women's residence access was undermined by the presence of dependent children. Returned adult children also represented a loss of remittances, an important source of non-agricultural income for the single women hearth-holds. This increased the vulnerability of the hearth-holds' livelihoods as remittances were used to purchase household goods and agricultural inputs.

7.3.2 Productive activities

The declining viability of agricultural activities means that hearth-holds have to diversify their livelihoods in order to sustain their livelihoods. The extent of diversification depends on individual women's portfolios. A pension is an income source for widows of former civil servants. Five single women in the sample of 22 received a pension from the estates of their deceased husbands. The money has been eroded by inflation. Three other widows in the sample of 22 could not access their deceased husbands' pensions because of lack of documentary proof of marriage. This was caused by illiteracy, the need to balance the allocation of time between the rural areas where the hearth-holds live and farm and urban areas where formal institutions of justice are located, limited access to financial capital needed for travelling and to pay the various fees required to acquire formally documented proof of marriage (Paradza, 2008).

Remittances are a source of income for elderly women with grown-up children. This was an important way in which hearth-holds that were based in communal areas could continue to receive financial income from urban areas after the loss of their husband who in most cases had remitted wages. Some of the elderly women who were raising children of employed parents also received remittances, but these were irregular and inadequate. Four of the women in the sample of 22 received a regular and reliable remittance amount. Seven of the single women had adult children in their homesteads who had failed to secure employment. The presence of adult dependent children in single women's homesteads highlighted in Table 7.1 indicates a failed opportunity for the woman to derive remittances from those children.

The contraction of urban employment opportunities has undermined hearth-holds' livelihoods. This is because with no employment the younger generations are forced to depend on the economic assets of older people. This has consequences for the welfare of the older generation as it undermines their economic welfare and increases pressure on their assets. The ownership of urban immovable property potentially enabled the hearth-holds to earn an income through rental of these properties. The income from these properties should have provided insurance to the single women in the form of rentals[67]. This was only possible if the women who had the properties had control of the properties and the rental income deriving from them. Four of the hearth- holds in the sample of 22 all, widows inherited urban immovable properties upon the death of their spouses. Two of the hearth-holds were receiving rental income. The other two were not, because their own children were residing in those properties, as shown in Table 7.1. These were adult children who, because they had failed to establish an independent dwelling in urban areas, relied on the family property. However, the livelihoods and housing problems of the younger generation pose problems for the older generation's potential to benefit from their ownership of urban housing (Paradza, 2009:423). The children and their families were not in a position to remit money to their mothers in the rural areas. The loss of this source of financial income increased the hearth-holds' livelihood vulnerability as they had to secure other sources of cash income in the communal areas. These examples illustrate how the negative impact of the economic situation in Zimbabwe disrupted the flow of financial capital through rural-urban connections. The high inflation and contracting economy resulted in the drying up of remittances and pension flows to the communal areas. This made the rural livelihoods more vulnerable to the deteriorating economic conditions as they lost their traditional sources of income that they used to invest in agricultural activities and develop and maintain their homesteads.

Legislation in Zimbabwe states that a man must provide for his legitimate and illegitimate children (Armstrong, 1998). This is to ensure that man bears responsibility for the care of his children. Although all the divorced women in the sample were aware of this provision, none of them was receiving regular alimony from their children's fathers. The constraints that influence the enforcement of Maintenance Law include the attitude towards child support influenced by customary law, allegations of women's abuse of the allowance, financial and practical problems, and fears of physical and other retributions (WLSA, 1992; Armstrong, 1998; Mupedziswa and Gumbo, 2001). The high unemployment among men partially explains the men's inability to pay maintenance for their children. In cases where these minor children are being fostered, some of the children lack formal documents to prove the paternity.

---

[67] To hedge against inflation, people were charging rent in foreign currency and/or demanding groceries (such as soap, sugar, cooking oil, maize seed, etc.) from the tenants.

Ten of the 22 single women in the sample were engaged in petty trading. The activity is popular because it requires a low initial capital outlay. The traders in this study can be divided into three groups according to their access to financial capital to invest in the venture. There were women involved in relatively skilled and capital intensive activities like cross-border trading. Women and men engaged in this activity needed to have secured a Zimbabwean passport, access to foreign currency and networks. The networks enabled the individuals to identify the goods for sale and the markets in the destination countries. The women and men normally sold artefacts to Mozambique, South Africa and Botswana. The second category of traders was involved in the sale of candles, salt, sugar, vegetables and second-hand clothing. The single women bought these commodities from the growth point, former commercial farms or urban areas. All the women engaged in this type of trading obtained their capital from relatives and friends. The availability of basic commodities is erratic because of poor transport and the traders' lack of capital to finance the activities. The business activity is also vulnerable to police raids and confiscation of commodities. This is because the women did not have trading licences. All four women regularly consumed their operating capital and constantly adapted the commodities in line with the availability and scarcity of commodities on the market. The commodity markets for vegetables were flooded during the peak season as there was an oversupply of cabbages and tomatoes. The single women consumed the excess stock and incurred losses. The women with only their labour to invest were involved in the trading of natural forest products, which they obtained from the bush and commercial farms. These included firewood, mice[68], tobacco, brooms, thatch and fruits. The women obtained the commodities free of charge, and the volumes were very low. The activities were time consuming. The returns were low for the labour involved. The availability of the commodities was seasonal. The productivity of these activities was declining because of the increased depletion of the forest resource. The natural produce which the hearth-holds exploited for sale was also vulnerable to destruction by bush fires during the dry season. Firewood sellers had to compete with young men who used scotch carts and produced in larger volumes than the head loading of women. The people who sold firewood and thatch stole these commodities from former commercial farms. The demand for thatch was low because of the depressed construction economy. The initiatives by people involved in trading to diversify into trading activities illustrates not only the importance of rural-urban connections in underpinning livelihoods but also the fact that in addition to people, money, goods and produce also flow between the two spatial divides. As the foregoing paragraph shows, the individual portfolios also determine the nature of the flows of goods and people.

[68] Mice are consumed as relish, that is, as an accompaniment to the local staple, *sadza* (maize porridge).

Chapter 4 highlighted the presence of a NORAD-funded dairy project in Ndamba. Members of the project were able to benefit from milk and ownership of cattle. This was a profitable product when it was available as there was high demand for milk. However, all except one woman in Ndamba had dropped out of the project because of high livestock mortality and the withdrawal of donor support.

Three single women brewed beer for sale in the villages and commercial farming areas. Two elderly single women manufactured pottery in Makuku and Ndamba, respectively. The women used clay, which they collected from the source[69] and transported by head, pounded, ground, sifted and used to mould the pots. There was a high demand for the pots from the local village and the Apostolic church[70]. The women bartered the pots for food and services such as draught, assistance with repair of the homestead and fencing. Both of the women had learnt the skill from their respective mothers. Pottery is a seasonal activity and requires a lot of firewood for the firing process. The sale of beer and pottery enables single women hearth-holds to raise financial capital in the rural areas. Some traditional activities of healing, death and birth rituals are almost exclusively performed by postmenopausal women. The high incidence of death, lack of effective health services, drought retrenchment and 'inexplicable misfortune' has meant that an increasing number of people turn to traditional means of solving their problems. This increased the opportunities for the hearth-holds engaged in healing, funeral rites and beer brewing to earn monetary income for their services.

With the exception of the traditional activities, women require a licence to practise. None of the single women in the sample had a licence. They all risked arrest by engaging in illegal activities without the required licences. The women who harvested natural produce and traded it were taking two risks. They stole the commodities from the commercial farms. They risked arrest by trading in these commodities without a trading licence.

Chapter 4 showed that even though some commercial farms had been taken over by the government during the fast track land reform, some European commercial farmers remained in Goromonzi. These farmers provided employment opportunities for people in the communal areas. Although commercial farm work opportunities exist in the study area, such work was poorly remunerated. Commercial farmers were failing to mobilise adequate labour from the communal areas. The farmers picked up the villagers at 04.00 hours to return at 15.00 hours. Six single women in the sample were employed on commercial farms. Two were permanent and the

[69] River bed at least five kilometres from the village.

[70] One Apostolic church insists that individual members keep 'holy water' blessed by the prophet on their domestic premises to protect them against illness and bad luck. The members are encouraged to store the water in clay pots.

rest temporary workers. Commercial farm work is hard and backbreaking with long hours and thus left the hearth-holds with no time to invest in their own subsistence land. This increased their vulnerability to crop failure and ends up increasing their dependence on the market for food and transfers. Furthermore, there is no job security in the commercial farm work because all the workers are laid off after the harvest. The hearth-holds complained that they no longer received benefits that they historically associated with commercial farm work, such as food subsidies from the farmers. Other commercial farmers used send their tractors into the communal area to plough for the individuals who worked for them while others provided monetary loans and subsidised the cost of their employees' children's schooling and provided housing. All the hearth-holds holds who engaged in commercial farm work were among the poorest households and had been engaged in this activity for all their lives.

Chapter two highlighted that the research area was in close proximity to Juru growth point. At the time of the research, the 60 shops at Juru Growth Point included butcheries (butchers' shops), bottle stores (off-licences), supermarkets, bars, nightclubs and petrol filling stations. Each of these enterprises employed an average of two single women. Business owners reported that the turnover of these shop assistant women was very high. An average worker stayed up to a year. Their main reason for leaving was pregnancy. The growth point harboured a substantial number of single women who claimed to be involved in selling food or working in the shops, bars and nightclubs. A focus group discussion with the women who were not employed showed that the women used to work in the shops, bars or nightclubs but that they left to start their own business. The business involved a day and a night job part. During the day, the women were engaged in cooking and selling food, hairdressing or vegetable vending. At night, they worked as prostitutes, which the women claimed was more profitable[71]. The customers were mainly truck drivers and members of the surrounding commercial, communal and mining community. The business owners provided the accommodation. The commercial sex workers have formed an association to work with the HIV/AIDS awareness groups. The majority of the membership was drawn from divorced, widowed and single younger women from the communal areas. These women indicated in the focus group meeting that they were not interested in subsistence agriculture as their work was more profitable and less prone to climatic vulnerability and the shortage of financial capital in the communal areas.

In Chapter 3, the communal area agriculture was described as entirely dependent on the rain. This is unlike the commercial farmers that had irrigation systems. The rainy season lasts from October to February which makes timing of planting very

[71] Though prostitution is an illegal and stigmatised activity in Zimbabwe, the women in the focus group freely admitted their involvement in the activity. However, It was not possible to pose questions about this sensitive issue to individual hearth-holds during the in depth interviews.

important. Even though the communal area commercial and subsistence agricultural production capacity was declining, the activity was still a part of the portfolio of some single women hearth-holds. Eight of the 22 hearth-holds in the sample did not engage in agricultural production at all in the 2004/5 agricultural season. The hearth-holds may have been minimising the risks associated with agriculture or lacked money, draft and labour to enable them to undertake agricultural activities. Chapter 6 showed that *matongo* enabled some single women to accumulate land in Makuku. These land-rich elderly women in Makuku leased out some of their agricultural land to land-short households and benefited from the harvest of their tenants. This enabled the hearth-holds to reduce their hearth-holds' exposure to the climatic variability. Fourteen of the 22 single women engaged in agriculture in 2004/5 farming season. Ten of the fourteen hearth-holds failed to produce enough food to meet their subsistence needs. The latter part of the chapter will show that these hearth-holds failed in their agricultural activities because they either lacked the agricultural inputs, draft power, labour or all three commodities to farm or their agricultural activities were disrupted by shocks like illness and mortality. Five of the 14 single women hearth-holds that engaged in agriculture produced enough for the subsistence of their hearth-holds. The households that performed relatively well in subsistence farming were those who purchased seed, hired or owned draught animals, mobilised additional labour and were able to fertilise their crop on time. They harvested at least 10 sacks of 50 kg of shelled maize each. The hearth-holds with access to regular remittances and or secured rental income from an urban property formed the core of the successful agricultural producers. This illustrates the importance of rural-urban connections in agricultural viability in the communal areas.

Draft power is important for preparing the land before planting. Although six of the 14 cultivating single women owned cattle, none of the hearth-holds owned a full span of traction animals. The livestock were inherited, acquired on the marriage of their daughters or procured through the NORAD-funded heifer project (see Section 4.7). Four of the single women (all widows) owned ploughs which they had inherited on the death of their spouse. However, bearing in mind that none of the single women possessed a span[72] of cattle, the ownership of a plough had limited impact on their ability to plough the land by themselves. The women who owned cattle pooled their animals for draught and also generated income from hiring out the cattle. Some single women received free ploughing provided by their relatives, male liaisons or their deceased spouse's relatives. However, free ploughing was done at a time and acreage convenient to the one giving the free service. This did not necessarily coincide with the recipient's plans and thus increased single women's vulnerability to climatic variability and the short agricultural season. The elderly single women who had access to money hired ploughing services. Others offered

[72] A standard span is four oxen (Wolmer and Scoones, 2000:587).

labour in exchange for draught. This reduced the amount of labour available for the timely preparation of their own hearth-hold's land and increased their vulnerability to crop failure. Three of the women used their hand hoes for tilling the soil, as they had no access to draught at all. The women who hand-hoed were in Ndamba village. The different ways in which hearth-holds with limited access to the agricultural inputs tried to organise alternatives illustrates the ways in which they attempted to mediate their vulnerability to subsistence agricultural failure. Here vulnerability helps us to understand how hearth-holds experienced threats and how they managed the threats to their livelihoods.

Vulnerability is caused by the HIV/AIDS pandemic, rapidly changing environment and market liberalisation policies. These have led to an increase in the number of single women hearth-holds and resulted in labour shortage among these hearth-holds. More than two thirds of the hearth-holds in the sample relied on their own labour for subsistence farming. The hearth-holds residing as lone residents on their homesteads relied on their own labour. This shows that the historical labour mobilisation through kinship and reciprocity either were no longer working or were not available to single women hearth-holds. This differential access to money for hiring labour increased differentiation between the hearth-holds, as hearth-holds that hired out their labour could not afford to hire in any labour. The two single women with means in the sub-sample obtained labour from the market. Nine of the hearth-holds had insufficient labour for their own produce because they hired out their labour to other hearth-holds and homesteads. They assisted with weeding, harvesting and processing maize in exchange for food and or cash. The five relatively successful hearth-holds were the main consumers of the hired labour. The healthy single resident hearth-holds all hired out their labour. Some hearth-holds were confined to the rural area because of their responsibility for small children. They too hired out their labour and thus reduced the labour available work their own land. This arrangement increased the vulnerability of the hearth-holds, which hired out their labour to food shortage. This was because they did not have enough time to invest in their own subsistence production. Vulnerability here refers to the exposure to shock and the difficulty with coping. The situation in which at least ten hearth-holds depended on five hearth-holds for labour shows the extent of dependency of the communal area population.

The government has in the past tried to reduce people's vulnerability to the short rainy season and poor soils by providing free seed and fertiliser to the rural inhabitants. The government seed was disbursed in February at the end of the rainy season. Some households try to use old seeds or buy from the market. In the 2004/5 season, the late disbursement of seed by the government increased people's vulnerability to crop failure. Single women hearth-holds were more vulnerable to crop failure as they were more reliant on the government seed and fertiliser. Ndamba villagers had an alternative seed source through their participation in the dairy and organic farming projects.

The low number of successful farmers in the sample shows the high risk of crop failure that hearth-holds that practise agriculture as a portfolio activity are exposed to. The hearth-holds who had a successful agricultural season reduce this risk by with holding all their produce from the market so they can use it as a buffer against crop failures in the future. The poor agricultural performance of single women hearth-holds begs the question of why people continue to invest in the activity when it is so risky. The stifling of trading opportunities by the government and the poor liquidity situation in communal areas has left subsistence agriculture as the only legal livelihood activity in the communal areas. This worked when the viability of agriculture was assured in the past through government subsidies and sustained flow of seeds, fertiliser and money to invest in agriculture from the waged workers. The people who engage in trading activities are liable to get arrested by the police as this is against the law to trade outside the prescribed commercial trading zones in the communal areas. Some authors are of the opinion that women continue to farm because they lack resources to diversify (Berkvens, 1997; Andersson, 1999; Huisman, 2005: 256). The limited capacity of people with no financial flows from urban areas to mobilise agricultural inputs illustrates the importance of conceptualising the rural and urban as a single social universe rather than two dichotomous spheres.

Single women hearth-holds attempt to mediate the risk of agricultural crop failure by working in vlei gardens. These are gardens that are located on wet lands, sponges which are less prone to the short rainy season than the arable lands which are solely reliant on rainwater. Garden produce provides a supplement to the staple diet and can be a source of income. Households without gardens provide a market (Bell and Hotchkiss 1991:210). Although they are less vulnerable to drought, vlei garden produce is vulnerable to wildlife, domestic livestock and theft. In order to reduce the loss of garden produce, individuals have to fence and secure their vlei gardens. The distance from the homesteads, decreased pasture and depletion of the forest resources made it difficult for the hearth-holds in this study to secure their garden adequately. As a result, their vlei gardens were under-utilised. The women in both villages had instead opted for home gardens. These are gardens located on the residential land, which are irrigated using domestic wastewater and water from the homestead well. The hearth-holds without a well and fence on their properties were not able to establish home gardens. Ndamba village established a community garden, which enabled the women to reduce the high risks associated with individual vlei gardens, seasonal water and domestic labour shortage. The Ndamba garden project is described below.

Ndamba's community-level initiative to reduce hearth-hold vulnerability to drought

In 1988, a co-operative garden was established in Ndamba village. This garden was located nearer to the homesteads than the vlei gardens. The site of the garden is land that belonged to the adjacent village of Mapfumo. The women's group identified the site as ideal for a garden and approached their *sabhuku* for assistance. The *sabhuku* arranged for a land swap with Mapfumo. The Ndamba *sabhuku* gave two homesteads in exchange for the garden site. The Ndamba *sabhuku* allocated the land to the women's group. A commercial farmer drilled two wells. A non-governmental organisation loaned them money for fencing. All women resident in Ndamba village at the time were eligible for co-operative membership. Initially they worked together, growing vegetables for sale collectively, hiring and organising transport to urban areas. With the proceeds, they paid off the fencing loan and realised some income. The co-operative benefited from the advice of all the NGOs that passed through the village. After some time, the women decided to subdivide the garden into individual plots within the garden. Each woman was allocated some land for individual use. The wells and fencing were still maintained communally. The garden gave the women opportunities to pool labour and to organise draught collectively. The availability of a fence and the community-level obligation to maintain it meant that the crops are more secure from theft, wild animals and livestock. During the rainy season when the hearth-hold labour has to be spread between the individual garden, arable field and co operative garden, the women abandoned their individual vlei gardens outside the co-operative gardens and focused on the land in the co-operative garden without compromising nutrition in the homesteads. Without the co-operative garden, the women would not have been able to produce supplementary food as the costs and labour of walking to, fencing and individually maintaining the vlei gardens are higher. The hearth-holds' participation in the garden project enabled them to receive inputs, information and advice that would be otherwise difficult to access at the individual household level. During the dry season when the water table in the wells was lower, the women took turns to water their vegetables, some watering in the morning and others in the evening to ensure there was enough water to go around. Members also assisted each other with weeding and watering vegetables in times of illness or when a member had to go away to attend a funeral or other social obligation. Portions of land were easily lent to other women by relatives or those not utilising their portion. The younger women who were not members of the co-operative benefited through borrowing the land belonging to the older women. One of the women was not using her plots because she was nursing an ill person. Two women did not use their plots because they had inadequate labour. Younger women borrowed plots from their relatives and mothers-in-law. The activity of gardening is most useful in enhancing household nutrition and decreasing expenditure on food. Two widows in the sample employed other people to work in their garden, though production is mostly for their own consumption.

This community-level initiative has reduced the vulnerability of women in the communal area to garden produce loss and rain dependent agriculture. Through this effort, women were able to secure access to land, water, fencing, market, technology and opportunities to diversify their livelihoods. Six of the women in the sample participated in the garden project. They were all in Ndamba. Makuku village attempted to establish a village garden as well. The village head allocated the co-operative some land for this purpose. However, the *sabhuku*'s mother claimed this land for herself as it had been allocated to her son who works as a teacher in town. As a result, Makuku had no co-operative garden. The minutes of the Makuku *sabhuku*'s secretary show that in Makuku 3 immigrants had set up gardens along the riverbanks and livestock routes to the river. This was another source of conflict between the autochthons and immigrants in Makuku.

7.4 Hearth-hold livelihood decision making: typologies of hearth-holds

An individual hearth-hold's livelihood assets and decision making determine livelihood outcomes. The life stories of single women hearth-holds in the sample were analysed using the pathways concept. The pathways of individual women are a matrix of different bundles of livelihood capitals and livelihood shocks. The pathways illustrate the variation in resources, contingency and agency of the individual single woman. Hearth-holds' health and/or burden of care and access to steady and reliable non-agricultural income are an important determinant of diversification capacity. The positions are also influenced by processes associated with the formation and development of households, together with the longer-term processes of accumulation and loss of assets (Francis, 2002:536). Because it is historical, the pathways concept allows us to take into consideration the impact of various processes on single women hearth-hold decision-making. The single women hearth-holds in the current study showed differences in level of diversification, depending on wealth, age group and location of single women-directed hearth-holds. Single women need access to living space from which they can organise access to opportunities to procure labour, land and financial capital from the market and/or networks. An examination of the individual women's pathways allows us to illustrate the diversity in hearth-hold decision-making.

The matrix of pathways shows that the individual hearth-holds studied were all unique. An attempt to draw generalisations revealed some common characteristics however. These are represented in Table 7.2. The typology divides hearth-holds into five groups according to their position on a continuum. These ranged from group I, hearth-holds that diversified into accumulation, to group V, hearth-holds that failed to construct viable livelihoods and succumbed to the shocks. The criteria used to classify the hearth-holds were the viability of their livelihoods, livelihood portfolios, and the capacity of the individual hearth-holds to sustain an independent

Table 7.2. Differentiation of single women hearth-holds.

Characteristics	Assets	Portfolio
Category I: Hearth-holds: diversification into accumulation of assets (4 hearth-holds)		
Independent, well maintained homestead, able to save and invest surplus		
Mean age of head (66) No longer child bearing Single for at least ten years Adult children Healthy No AIDS-related burden	Healthy head Independent, well maintained homestead Agricultural land Co-resident with family members (labour) livestock Owned agricultural inputs (draught, seed, labour), livestock Kinship, Social networks	Diverse viable activities, regular receipts of non-agricultural income Surplus agricultural production requirements High capital-high return activities (livestock, pig rearing, beer brewing construction, cross border trading, pottery) Land baroness
Category II: Hearth-holds that are managing (3 hearth-holds)		
Independent homesteads. Get enough to meet day to day needs. Do not accumulate		
Mean age of head (59) No longer child bearing Small resident population (max two residents) Healthy Cannot afford to pay for medical care No burden of caring for ill and/or young children	Healthy individuals Independent, poorly maintained homestead Own labour, community, arable land, Social networks	Erratic remittances High reliance on transfers from government and non-governmental agencies for agricultural inputs increases risk of agricultural failure Manage risk by trading their labour in exchange for food and agricultural inputs Depend on NGO activities Pottery, beer brewing, healing
Category III: Vulnerable (7 hearth-holds)		
Independent homesteads poorly maintained who rely heavily on transfers		
Mean age of head (54) Child bearing and post-child bearing Hosting ill adult and/or young children Cannot afford medical care Losing assets over time	Independent homestead, poorly maintained, arable land Community membership Heavy burden of care reduces capacity to pursue livelihood activities	Cannot produce subsistence even if they receive all agricultural inputs on time Rely on transfers and charity (family, church, state) Illegal trading of poached natural produce Commercial farm work

Table 7.2. Continued.

Characteristics	Assets	Portfolio
Category IV: Dependent vulnerable hearth-holds (7 hearth-holds)		
Hearth-holds embedded in other people's homesteads lack livelihood capitals to negotiate access to independent homestead and sustain an independent livelihood		
Mean age of head (31) Child-bearing and/or ill Dependent children Hearth-hold split between two residences	Lack independent homestead Labour and sometimes Social networks Not accumulating assets to buffer hearth-holds in the short term	Lack livelihood capitals to negotiate access to independent homestead High reliance on host (parents, family) Experimental trading; irregular, erratic, low income; high risk-low return activities
Category V: Succumbed hearth-holds (1 hearth-hold)		
Wholly dependent for food, shelter and care		
Age of head 32 year old All members ill Dead, dissolved	Kinship support	

Source: Author Fieldwork 2004/5.

homestead. The typologies are not fixed. Though generally hearth-holds gradually slipped downwards as they struggled with the increasingly difficult conditions, some progressed to less vulnerable categories over time. Age is important because it relates to experience, accumulation of assets, and possible assistance from children. The hearth-holds' relationship to the village head was also an important asset for their hearth-holds in negotiating access to other livelihood capitals in the village. The extent to which hearth-holds were able to gain access to livelihood capitals and mediate vulnerability to the volatile and deteriorating environment over the life course determined the sustainability of their livelihoods.

7.4.1 Hearth-holds type I: diversification into accumulation

Type I hearth-holds all headed and controlled independent, well-maintained homesteads. The average age of the hearth-hold head was 66 years. Over time, all the hearth-holds had successfully diversified their livelihood portfolios, which reduced their vulnerability to agricultural failure and poor economic conditions and secured their livelihoods. The hearth-holds had diverse non-agricultural income sources in their portfolios. The minimum age of the hearth-hold head in this category was 50, meaning that the women had acquired experience and accumulated assets, and that most likely their children had grown and left the house. All hearth-holds heads in this category were relatively healthy and were not hosting ill adults. Though some of the hearth-holds had lost family members to AIDS, the accumulated asset base had enabled them to weather the setback. The hearth-holds in category I formed the core of the successful agricultural producers. They were able to mobilise agricultural inputs for their hearth-holds well before the onset of the rains and could produce surplus from their agricultural activities. They had stopped marketing their agricultural surplus, using it instead as a buffer and for negotiating access to labour and other livelihood capitals. They hired labour from the poorer hearth-holds. The women were able to accumulate some savings, as shown by the capital intensive investments they made. Diversification provided a buffer against short, medium and long-term risks. The hearth-holds' livelihoods remained vulnerable to the volatile conditions in different ways, as highlighted in the following case summaries.

Mary: weathering shocks through diversification

Mary, who was living in Ndamba village, was a 50-year-old widow heading a homestead and already widowed for 12 years in 2005. She had the best-developed and maintained homestead in the two villages of Ndamba and Makuku. She lived at the house with her two adult sons and a grandson. One of the sons was married and his wife lived here with Mary. Mary and her husband were allocated the homestead by *sabhuku* Ndamba in 1977. The husband worked in the urban areas while Mary looked after the homestead and farmed, producing maize for sale. The husband constructed two buildings. In 1990, Mary was elected to go and represent the village at a meeting of co-operatives in Tanzania. She obtained a passport to undertake this trip. Her husband died in 1993. Mary continued to live and farm on the homestead after her husband's death. She also had access to the rental income from the property in Harare and her deceased husband's pension. Mary managed to put her children through school. In 1998, she decided to use her passport to try cross-border trading. She received assistance from her neighbours at the house in Harare[73] who had relevant experience. Mary used her savings from crop sales to invest in the cross-border activities. She bought artwork from Zimbabwe for resale in South Africa and brought blankets and clothing for resale to people working in the mines. She added another building to the homestead and invested in solar power. She also completed the construction of the house in Harare, which she and her husband had started while he was still alive. Mary's married son lost his employment

[73] Mary interacted with her neighbours when she went to Harare to visit her husband who had lived and worked in Harare.

during operation Restore Order and moved back to Mary's homestead with his wife. Mary's unemployed son and the daughter-in-law helped her to keep the house and looked after it when she was absent. The widow's 19-year-old son, who was also unemployed, lived at the homestead. Mary did not travel between November 2004 and February 2005 as she was engaged in agricultural production on her homestead. The widow obtained all her agricultural inputs from the market and well before the onset of the rains. In 2004/5 she harvested 19 90kg bags of maize. She had 13 bags left over from the previous year in her granary. She did not sell any of her maize. She hired people from the village for agricultural work, fetching firewood and working on her co-operative garden. Mary's elder son had some erratic employment and used the proceeds to fund his homestead development. The younger son did not engage in agricultural activities. When foreign currency became scarce as a result of the sanctions imposed on the Zimbabwe government in 2001, Mary resorted to the black market. During operation Restore Order, the artwork suppliers relocated and controls at the border increased. The widow went to South Africa but returned with nothing because she could not afford to pay the high import duties levied by the customs and excise officials at the border.

Mary's hearth-hold weathered the shock of losing her spouse with the assets that they had accumulated over time. She continued to farm and managed to educate her children. When her son lost his employment, Mary's hearth-hold's diversifed portfolio limited its exposure to this shock. Mary's timely decision to venture into cross-border trading reduced her hearth-hold's vulnerability to agriculture-related activities. As a result, she consolidated her portfolio by investing in immovable property. Mary's access to foreign currency shielded her homestead from the inflationary pressures that were haunting the Zimbabwean economy. The hearth-hold's livelihood portfolio remained vulnerable to the licensing regulations in Zimbabwe, which constrained her business, and the scarcity of foreign currency. Mary risked arrest by trading without a licence and by acquiring foreign currency on the black market. The widow reduced these risks by continued agricultural activities. As a result, she had a store of maize which she could consume or dispose of in times of stress. The availability of resident adults in the house, caused by a delayed departure of adult children from the parental home, increased the woman's access to labour. The son's wife helped with the agricultural activities and the children provided security for the homestead when Mary travelled on business. She was absent for periods ranging from 2 to four weeks at a time. The adult children's livelihoods and those of their families are vulnerable to the death or illness of the mother or the collapse of her business in the long term, although the widow has insured against this risk by investing in the urban property. By engaging in cross border trade and renting out the urban property. The example illustrates how a diversified portfolio, strong economic rural-urban connections and strategic decision-making enable a hearth-hold to survive the loss of a household head and maintain assets in a hostile environment.

Ann: combining experience and seniority to consolidate a homestead

Ann was approximately 80 years old in 2005. She lived in Makuku. She had never been to school[74]. Ann lived in a homestead which was well fenced and had an independent well and toilet. There was a chicken coop and a kraal for her six head of cattle. The homestead had one brick and asbestos structure, a granary and a thatched hut. There was a kitchen garden on the property. In 2005, Ann lived with her divorced daughter and divorced son (a war veteran[75]) and his two daughters. The divorced son and daughter had relocated to their mother's homestead at the end of their own marriages. The divorced son left his own homestead as a *dongo* in Makuku village. The son had purchased this land on the market as he was not in his patriarchal village. Ann was born in Makuku and left the village when she married. Ann's husband was a police officer in the urban area and she was farming in the rural area. This arrangement enabled the couple to secure financial capital to invest in their agricultural activities and reduce their vulnerability to the short agricultural season, frequent drought and crop failure. They had a son. The husband died after a short illness Ann decided against a *kugarwa nhaka* marriage and moved back to her natal village in Makuku. This was because the deceased husband's brother already had two wives and Ann decided that joining his homestead would have reduced her hearth-hold to poverty. There, her father, who was the *sabhuku*, allocated Ann her own homestead and helped her to construct a dwelling and clear the land for farming. Ann received her deceased husband's pension, worked on commercial farms, performed *maricho*[76] for other villagers and moulded bricks for people in the village to sustain her hearth-hold. At the time Ann was doing this, the Zimbabwe economy was thriving, employment was easy to come by and Africans were not allowed to own urban property. They invested the proceeds of their employment into the development of rural homesteads for their retirement. This increase the demand for bricks and Ann was able to secure financial income from brick moulding. Through these efforts, she developed her homestead and acquired livestock. Ann

[74] Ann has no concept of dates and time. She could not even relate to major events like ESAP, independence, etc.

[75] The divorced son did no agricultural work. He received the monthly gratuity for war veterans from the Government of Zimbabwe but he did not contribute to the house. He drank away all his money and did not even pay his daughters' school fees.

[76] Piece work performed for other villagers in exchange for food, soap, salt, old clothing or money.

left her homestead in her father's village when she married another man. They had two children, after which the man died. Ann returned to her homestead in Makuku. When she was too old to work on the commercial farms, Ann focused on farming her own land and brewing beer. This diversification reduced her vulnerability to poverty and old age. In addition to her own land, Ann controlled the *dongo* (see Section 6.2) of her other son who had moved to a former commercial farm, a *dongo* which belonged to her now deceased brother and her deceased parents' *dongo.* She arguably had the most land in the village. Ann leased out some of her land to land-short immigrants in Makuku. Although she had six head of cattle, four of them were calves so she had to enter into some pooling arrangements to plough her land. She purchased agricultural inputs on the market and did not rely on the government-supplied inputs. In 2004/5 she harvested 3 50 kg bags of maize. Ann bought firewood from the young men in the village. The *sabhuku* Makuku tried to sell some of her arable land but she successfully resisted this move. She was worried about the increasing number of people settling in the village as this increased pressure on her accumulated land holdings. Every week she brewed beer for sale. Ann's son on the commercial farms supplied her with maize and firewood. When her cows calved, Ann also sold milk. This increased the diversity in her activity portfolio.

Ann's pathway shows how a hearth-hold over time moved from a dependent vulnerable position (Category IV) to one where she was able to accumulate assets and successfully diversify (Category I). The woman's long experience in brewing beer enabled her to establish a reputation, which ensured a constant stream of customers to her beer parties. The woman's senior position enabled her to accumulate land in a village where land holdings sizes were increasingly falling because of the land market. This accumulated land was a source of livelihood as she hired it out to land-short people. The woman's decision to accumulate land at a time of increasing land demand placed her in a good position and increased her security, as land was becoming a scarce commodity. Ann's senior position enabled her to defend her land against the *sabhuku,* who had a lower rank. The presence of a war veteran son on her homestead protected her land interests against the *sabhuku.* The return of her adult son and daughter to the homestead following the end of their marriages increased her access to labour for beer brewing and agriculture. The woman also received remittances from her other son and these enabled her to gain maximum returns from her agricultural investment. The consumption sheet for this widow showed that, between April and August 2005, she was purchasing sugar, yeast, grain and firewood for brewing beer on a weekly basis. The woman's successful diversification into beer brewing and 'estate management' enabled her to overcome the risks of the agricultural decline. The ownership of cattle and control of land represents a form of savings, which she could liquidate in times of stress. Ann's health despite her advanced age was another important asset. The hearth-hold developed her homestead through harnessing people's investment into rural property. When this was no longer viable, Ann diversified into beer brewing, 'estate management' and the receipt of remittances from her employed son. This ensured the hearth-hold's continued access to goods and services on the rural-urban continuum, goods and services that were crucial for the maintenance of the homestead.

Matty: immigrant widow consolidating a successful homestead

Matty was aged between 65 and 70 years. She lived in Makuku village with two orphaned grandchildren. The house was a three-bedroom brick under tile structure built by her husband when he was alive. The property was well fenced and maintained. There was a good orchard, a well-maintained thatched kitchen, protected well; cattle pen and chicken coop. Matty purchased all her agricultural inputs and hired labour and draught. She harvested 750 kg bags of maize in 2004/5. Matty lost a 32-year-old widowed daughter, a 26-year-old married daughter and her husband between 2000 and 2005. She had a house in Harare and her only surviving child, a divorced daughter, lived in the United Kingdom. This daughter was a source of remittances. Matty had one cow on the premises, two goats and 28 chickens. Matty came to live in Makuku in 1976 after her husband retired from his job in town. The couple retired to the rural areas where life was cheaper and to free up the urban property so they could rent it out for a fee. The income from the urban property would be the source of financial capital for the couple. The couple purchased the land from the *sabhuku*. The husband had until then been employed as a factory worker in Harare. Matty's husband died after a short illness. Matty succeeded to her husband's estate as all his male relatives had died before him. She also inherited a house in Harare, a plough, a scotch cart and a bicycle from the deceased husband's estate. She collected rent from the house in Harare and received remittances from her daughter in the United Kingdom. Matty hired labour for the collection of firewood and the maintenance of her homestead garden. When ill, Matty travelled to her daughter in the UK for medical treatment.

Matty's hearth-hold moved to Makuku early on in life, so that by the time the multiple shocks of ESAP, operation Restore Order and economic decline occurred, her hearth-hold was well enough established to weather the shocks because of the constant receipt of remittances, access to her deceased husband's pension and rental from the house in Harare. She was so well established that, even when the hearth-hold suffered multiple adult deaths in a short space of time, the remaining income sources were enough to sustain her hearth-hold at above subsistence level. Matty was able to survive in Makuku as an immigrant because she obtained services from the market. Her access to foreign currency from her daughter in the United Kingdom buffered her hearth-hold against inflation. Matty's hearth-hold's connections changed over her life time in line with their changing circumstances. As with all the hearth-holds in category I, the homestead was developed through the exchange of labour and finance between the rural and urban spaces. The rural area served as a retirement home for the couple while the money earned in the urban areas was used to develop the homestead and accumulate agricultural assets. Though the hearth-holds lost human capital through the deaths of family members and the head of the house, the hearth-holds secured financial income by maintaining the ownership of an urban property and relationship with a daughter in the United Kingdom that ensured the hearth-hold's access to financial capital. This reduced Matty's hearth-hold's vulnerability to the depletion of human capital, the worsening economic environment and the increasing difficult conditions under which the contribution of agriculture to the livelihood portfolio declined.

Riba: consolidating a livelihood through opportunistic decision making

Riba was 70 years old in 2005. The Makuku *sabhuku* was her son. Riba lived with a 36-year-old daughter who had six children[77]. Riba was responsible for her grandchildren's school fees, clothing and care. Riba's other children, two sons and a daughter worked and lived in the urban areas. They sent her remittances. She had two employees aged 11 and 14, respectively. One herded the livestock and the other helped her with the housework and care of her grandchildren. She had a cow, six pigs and some chickens. Her homestead was well developed. She had a large piece of arable land and had also accumulated control of more land through upward inheritance (see Section 6.3.1) and by assuming control of land belonging to her son who had moved to live at his place of employment. The son who was a teacher lived at the accommodation provided for teachers at the school. During the school holidays he came to the village and lived in his mother's homestead. Though he had been working for more than five years, the son had not accumulated adequate capital to develop his homestead in the rural areas. This means without his mother's homestead, he had no place to retire to at the end of his working life. The diverse portfolio enabled Riba to accumulate financial capital which she invested in physical assets. In 2005, she completed the construction of a three-bedroom brick under asbestos house and repaired the roof of her thatched kitchen. Riba's husband worked and lived in town while she lived in the rural areas and farmed. When her husband died, Riba continued to live in Makuku village. She inherited a house in Harare and continued to use the rural homestead. She also continued to farm and was producing surplus for sale. Through these efforts and her husband's pension, Riba managed to educate all her children. In 2001, Riba abandoned her homestead in Makuku after the villagers accused her of witchcraft[78]. She sold her seven cows and returned to her natal village in another communal area in Zimbabwe. She lived there for two years. (While in her natal village, she was using her deceased parents'

[77] This daughter was also one of the hearth-holds in the sample though she was in category III.
[78] One of her grandchildren died. The villagers hired a faith healer who accused Riba of killing the child.

dongo.) Riba returned to Makuku to take care of her adult son who was now ill and living on his own. Her son died in 2004 from tuberculosis and chest pains. In 2003, she sold some of her arable land to immigrants in Makuku and used the proceeds to invest in a pig-rearing venture. Riba's was one of the few homesteads that purchased agricultural inputs on the market and she used all her arable land during the 2004/5 agricultural season. She also collected government seed but put it away for the following year as it was delivered late. Riba's grandchildren and labourers provided additional labour. Her children also gave her money to hire extra labour (*maricho*). She had vegetables in the vlei garden. In the 2004/5 season she harvested a surplus. She bought more grain on the market, however, to insure her hearth-hold in case there was another drought. She built a new three-roomed house with the assistance of her children, and this was completed in May 2005. Riba's son-in-law delivered asbestos roofing. In June 2005 she collected thatch to repair her old home. She hired some boys to cut wood for her and some poles to reconstruct the pigpen. During the dry season, Riba produced pottery pots. She hired labour to collect clay and firewood for her pottery business. The demand for the pottery was high and Riba was having difficulty meeting the demand. Riba usually bartered pottery for grain. Although she has an urban house, she does not receive rent from it as one of her sons resides in it. In September 2005, she had to go and care for an adult daughter who was ill. Riba's pottery business suffered in her absence.

Riba's life story tells of a hearth-hold that survived through a series of livelihood shocks, including the death of the male head, an adult son and the temporary relocation of the hearth-hold's head. This is another homestead, which had accumulated enough assets to tide it over the shock of losing the homestead male head. Riba's pottery skills have proved a solid asset against declining returns to agriculture and the deteriorating economic conditions in Zimbabwe. Riba used her pottery to accumulate grain. Riba's residence in Makuku and her position as the *sabhuku*'s mother enabled her to take advantage of the land market. She disposed of some land to raise money for a pig project. This challenges the general belief that single women hearth-holds are just care takers on inherited land as in this case, they even have the power to dispose of the inherited land. Riba also leases out some of her accumulated land. Although she suffered a setback when she abandoned her homestead, she was able to rebuild her livelihood when she returned because of her monopoly of the pottery market and, accumulated land interests and the income from the urban property which then was rented out to tenants. Like Mary, Riba's portfolio was robust enough to withstand the death of an adult son and the burden of caring for her daughter and her six children. She coped with the demands of looking after small children by hiring a maid. Her position as the *sabhuku*'s relative made it easier for her to resist any move to appropriate her land for the market. Riba's access to cash enabled her to pay for services like pottery clay collection and firewood collection. The hearth-hold's portfolio shows that the unit was accumulating assets and reducing its dependence on risky agricultural activities. The type of investments she made shows the extent of her hearth-hold's liquidity in a highly unstable environment. Riba's life history shows how hearth-holds manage the loss of a bread winner and cash income from wage labour by increasing dependence on remittances and developing a market niche in this case the pottery business which enabled her to continue to tap into financial flows on the rural-urban connections.

7.4.2 Hearth-holds II: managing

Hearth-holds in Category II had an average age of 59 years. The three hearth-holds (two in Ndamba and one in Makuku) in this category were managing. These hearth-holds managed through their diversified activities to meet their daily needs. They however remained vulnerable to death, illness and climatic variability. However, category II hearth-holds were poorer than hearth-holds in category one as they lacked the livelihood capitals to maintain their homesteads, diversify into less risky activities and buffer them against a catastrophe. Although type II hearth-holds managed independent homesteads like those of type I their homestead infrastructure was deteriorating over time. Type II hearth-holds had a small resident population. Besides the homesteads, their assets were their health and labour. These hearth- holds had limited access to non-agricultural income which undermined their capacity to engage in agricultural activities and diversify their asset portfolios. The hearth-holds relied on government seed handouts, free ploughing provided by their relatives, and their own labour. Over time, the hearth-holds lost assets due to the recurring shocks caused by HIV and AIDS, the volatile socio-economic conditions and Operation Restore Order. Unlike the hearth-holds in Category I, type II hearth-holds were not able to restore their livelihoods after suffering a setback. Women in these hearth-holds offered labour in exchange for food, which reduced the labour available for the timely preparation of their own land and increased their vulnerability to crop failure. These hearth-holds compensated the diminishing returns of agriculture by diversifying into less risky activities in the locality like working for better-off households. The diversification strategies they followed ensured that they maintained the viability of their hearth-holds on their independent homesteads. They did not have the burden of caring for ill adults or young children. However, the hearth-holds were highly dependent on their capacity to perform work, and ill health or the assumption of a burden of care would push these hearth-holds into Category III.

Petronella: managing agricultural risk through selling one's own labour

Petronella was a 57-year-old illiterate widow who was born in Ndamba. The widow lived with a 19 year-old unmarried son on the homestead. The well and toilet collapsed and the buildings were in desperate need of maintenance. This rendered the hearth-hold vulnerable to the loss of physical capital and shelter. Petronella had two head of cattle, which she obtained through the Heifer project in Ndamba. She started working on the commercial farms as a teenager. There she met and married a Mozambican. They lived and worked on the commercial farm where they were provided with housing, food subsidies and access to a clinic by the farmer. The benefits from the employer reduced the household's vulnerability to disease and poverty. The woman and her husband left the farm during the liberation war in Zimbabwe and secured a homestead in Ndamba with their four children[79]. The couple continued to work on the commercial farms. They never owned any livestock and did not engage in any subsistence agriculture. They relied on the commercial farms for their entire livelihood. During the agricultural boom in the early 1980s, the couple obtained subsidised draught power, seed and fertiliser, and engaged in agriculture on their homestead. They temporarily abandoned commercial farm work as they harvested surplus produce which they sold, and even constructed a granary. The husband resumed work on the commercial farms and Petronella diversified into local piecework. The husband fell ill and died from chest pains in 1998. Petronella assumed control of the homestead and remained in Ndamba. Her three children stopped attending school after the death of their father, as she could not afford to pay for them. Petronella's children married and moved out. Petronella survived entirely on selling her labour. This left her with little time to invest in her own arable land. In the 2004/5 agricultural season, she tried to engage in subsistence farming. She used old seed and government supplied seed, which was delivered late. She did not apply any fertiliser to her crop. The woman hired someone to plough her arable land but she was not able to pay him. She harvested less than one 20 kg bag of maize for the 2004/5 season. This was inadequate for her hearth-hold's consumption needs and rendered them vulnerable to starvation. In order to mediate the vulnerability, Petronella abandoned her co-operative and garden plot and focused instead on working for those households who could afford to buy agricultural inputs on time and hire labour. She provided labour on a more or less continuous basis for households in Category I. She worked in other people's arable fields and vegetable gardens, and even collected firewood for them. She was paid for the work in maize meal grain on which she and her hearth-hold relied for

[79] Apparently, the guerrillas threatened to kill any *sabhuku* who refused to allocate land to other landless Africans.

domestic consumption. The widow also did the laundry at the local school and was employed to sweep the dairy. She occasionally brewed beer when she had accumulated cash for all the inputs. She also provided healing services to the local community[80]. When the cows calved, she had milk for home consumption and for sale to other villagers. One of her cows died in August 2005.

The hearth-hold's pathway shows how a livelihood portfolio shrinks over time as a result of its vulnerability to death and the variable climate and strategic decisions that sustain the hearth-holds and enable it to maintain a livelihood. Through Petronella's pathways we reconstructed a marginal livelihood which is sustained by opportunistic decision making. The move to Ndamba and diversification from sole reliance on the commercial farm enabled the hearth-hold to survive the death of the husband. In Ndamba, the widow had access to an independent homestead which was not vulnerable to the fast track land reform that undermined commercial farm-based livelihoods. The decision to relocate introduced her to the network through which she gained access to employment and to NGO activities through which she managed to diversify her portfolio further. The portfolio of May 2005 shows that the hearth-hold was constructing a livelihood exclusively from non-farm work which involved performing agricultural chores such as harvesting, shelling and winnowing maize; fetching firewood; herding cattle; working in the co-operative garden; working on the commercial farms; cleaning the heifer project premises; and doing laundry for schoolchildren. The woman was paid in cash and/or produce for her work. The resident son added to the livelihood by herding cattle and selling firewood. The Ndamba village head's wife facilitated the widow's access to the school where she was doing laundry. Petronella's participation in the heifer project earned her a job as a cleaner at the dairy project. The performance of continuous piecework requires reliability and availability to provide a service to clients because some tasks such as gardening, herding cattle, sweeping at the dairy and laundry are routine and can only be performed by a healthy woman without maternal responsibilities. Although this left her with inadequate time to invest in her own subsistence farming, the activities reduced her hearth-hold's direct exposure to agricultural failure and poverty. Petronella's hearth-hold life story explores the circumstance of a communal area resident that had limited access to financial capital. This was mainly because the hearth-hold failed to export labour to the urban areas. Though the hearth-hold lacked direct rural-urban connections, she indirectly benefited from the flow of good and

[80] In August 2005, a young woman who was pregnant died in Ndamba village. According to custom, the baby and mother are not related in afterlife. So they must be buried separately. Previously, the hospitals would remove the foetus. In this case it was not done. The dead girl's grandmother insisted that they take out the stranger in her grand daughter's womb. She threatened that, if they did not do so, anyone else who fell pregnant in the village would die in the sixth month like her grand daughter. Ndamba village contracted Petronella to separate the corpses. The foetus was buried in the riverbed, as is local custom. The dead mother was buried on an anthill, like all adults. Petronella was paid with a goat, a chicken and a million Zimbabwe dollars.

services by developing labour contracts with the hearth-holds that had direct access to the rural-urban nexus. The hearth-hold's portfolio of activities, though diverse, only enabled her to meet her daily needs. The livestock show how external interventions by institutions like the government and non-governmental organisations can increase the people's livelihood vulnerability by increasing their access to livelihood capitals in this case livestock, which enable the vulnerable people to diversify their activities and acquire assets that can enable them to mobilise other capitals.

Tracy: exploiting a niche to secure livelihood

Tracy was a 68-year-old illiterate woman living alone on her parents' *dongo* in Ndamba. There was one hut. She had no well or toilet and relied on her neighbours' facilities. The *sabhuku* was her nephew. When she was young, Tracy married a man from another communal area. Tracy and her husband lived in Harare and worked as domestic workers. The husband died before the couple had set up a homestead. They had two children. Tracy married her deceased husband's younger brother and had more children. When her second husband died, she left the village and her children after the husband's family accused her of witchcraft. She returned to Ndamba, her natal village. She lived with her parents and learnt pottery skills from her mother. When her children had grown up and were working, they bought a homestead for Tracy in another communal area,[81] and she moved there. She lived for seven years farming and making clay pots. She did not get on with her daughters-in-law. Then she was called back to Ndamba to look after her widowed sister who was ill. When the sister died, Tracy stayed on in Ndamba. The children did not visit or send any money to her because they accused her of witchcraft. Her nephew, the *sabhuku*, ploughed some land for her for free in 2004/5. She hoed the rest of the land by hand. She planted organic seed that was disbursed free of charge by the NGO, Fambidzanai. She gathered organic fertiliser for the seed. She also planted old seed. She did not have any chemical fertiliser. She did not have a plot in the co-operative garden as it was established in her absence. She borrowed land from the *sabhuku*'s wife where she grew some vegetables for herself. She harvested less than 5 kg of maize in 2004/5. As this was inadequate to feed herself, she took up casual work for a local woman. The widow also made clay pots. The pots were always in demand as they were used as domestic utensils and increasingly in the church. She bartered clay pots for maize and then sold the maize for money. She was the sister of the spirit medium. She had the 'job' of brewing beer for the various rituals at the spirit medium's homestead. Beer takes two weeks to prepare and brew. During that

[81] A woman who is in the care of her natal family is perceived as a dependent of the natal family. Such arrangements also imply failure by the husband and/or children to look after their mother. Tracy's children wished to fulfil their obligation to look after their mother in her old age. Since she had lost the homestead in their father's village, the children bought her a homestead on the informal market in another communal area.

time, she abandoned her homestead and resided at the spirit medium's homestead. The spirit medium provided food for her. The frequency of ceremonies ensured that the widow was almost always working. When she was not brewing beer, she sold her pottery. She collected pottery clay by head from the riverbed and relied on the Ndamba forest for firewood. Tracy relied on the Ndamba *sabhuku*'s children to assist her with pricing her pots and looking after her homestead when she was absent. Tracy also regularly received drought relief food from the government.

Tracy's current situation emerged out of frequent dislocation as a result of which she was not able to accumulate livelihood assets like the women in Category I. However, she confronted each shock with the livelihood capitals at her disposal. The strategies included *kugarwa nhaka* (see Box 5.1) and *matongo*. Tracy's relocation to Ndamba after the death of her second husband gave her an opportunity to learn pottery skills from her mother. Tracy's obligation to care for her ill sister forced her to abandon her homestead. In that move, she lost a homestead but gained another through *matongo*. Tracy settled in Ndamba and managed to construct her livelihood through kinship, agricultural activities, and her pottery and her beer brewing skills. Kinship enabled her to access *matongo,* the Ndamba community and the spirit medium. The pottery making and beer brewing were exclusive skills through which she negotiated access to finance and labour. Pottery is seasonal. However, work at the spirit medium's house continues throughout the whole year. Tracy farmed because she had access to land, and the Ndamba community enabled her to access agricultural inputs through the NGO activities. Even so, she failed to produce enough for her own subsistence. This rendered her hearth-hold vulnerable to crop failure and starvation. Tracy mediated this vulnerability by working for the spirit medium and bartering her pottery for grain. Tracy's portfolio ensured her survival even when her children withheld their support. Tracy's livelihood was vulnerable to her falling ill and or succumbing to old age at the time of the research.

Tessa: from farm worker to subsistence farmer

Tessa was a 52-year-old widow who resided in Makuku. She lived with her 18-year-old son and a five-year-old nephew. The house was a well-constructed but poorly maintained brick structure. There was a well, which had collapsed. Tessa grew up in Makuku village. She married a commercial farm worker. In 1989, Tessa and her husband negotiated with her father who was the *sabhuku* in Makuku for space to construct their residence. They constructed two huts. They had three children. Tessa and her husband lived in Makuku and full-time commercial farm workers. They obtained food from the commercial farming areas. Tessa's husband fell ill and died in 1995. Over the years, Tessa's huts started to disintegrate and she could not maintain them. Tessa negotiated with the *sabhuku* for permission to occupy a *dongo* that belonged to a deceased sibling of Tessa and the *sabhuku.* The deceased man had constructed a solid brick and mortar house. His wife and young children left the village after his death. Tessa and her sons moved into the 'new' house and had better shelter. Tessa maintained her own *dongo* and used it for agriculture. Tessa's sons were reluctant to invest in either of their two *matongo* because they feared the risk of eviction by their mother's relatives. Tessa's health was deteriorating as she became older. In 2000, Tessa stopped going to work on the commercial farms and focused on farming in Makuku and leasing out land to immigrants. The hearth-hold had inadequate labour and relied on government seed for planting. Tessa's *sabhuku* brother sometimes ploughed for her. One of Tessa's employed sons sometimes sent her remittances. Tessa performed *maricho* for other households in Makuku in exchange for grain. During the fieldwork period, Tessa's 18-year-old son secured employment in the urban area and moved out of his mother's house. However, he did not send any money to his mother in Makuku.

Tessa's is a hearth-hold that, after working all her life, did not manage to accumulate anything. This is consistent with the finding (Sachikonye, 2003; Rutherford, 2001) that commercial farm workers receive the lowest wages. The family managed to negotiate homestead access through Tessa's relatives in Makuku. As a result, their status was better than that of immigrants. Tessa managed her failing physical health by withdrawing from the commercial farms and focusing on the village-based opportunity. However, she still needed to produce her own food. When she failed to do so, she coped by exchanging her labour for food. Tessa negotiated her way round her failing homestead infrastructure by moving to a better-constructed *dongo*, with the consent of her brother, the *sabhuku*. Tessa's sons were vulnerable to eviction from the village as immigrants. This discouraged them from maintaining their mother's housing. Tessa's story illustrates the untenable situation of hearth-holds that failed to maintain the flow of goods between rural and urban areas. They end up with limited financial capital to invest in the development of their rural homestead or educate their children, which also undermines their access to remittances and waged work. Tessa struggles to piece together a livelihood with very limited resources and as a result fails to maintain an independent homestead without the reliance on regular transfers from her kin in the village. The case also highlights how *matongo* mediate hearth-holds' shelter vulnerability by providing alternative accommodation to those who fail to establish and/or maintain their own homesteads.

7.4.3 Hearth-holds III: vulnerable hearth-holds

The women in this category lived independently on *matongo*. The mean age of the hearth-hold head was (54). These hearth-holds depended for most part of their income on transfers from the State, church, or family. Type III hearth-holds could not produce subsistence, even when there were good rains. Type III hearth-hold heads were either ill and/or had a relatively heavy burden of care, as over time they became ill, were hosting and taking care of an ill adult, and/or were taking care of small children. The burden of care limited their capacity to undertake livelihood activities. They only invested in agriculture because they obtained the agricultural inputs from the State or through charity from other homesteads. Some of the hearth-holds participated in the opportunistic trade of natural forest produce, or did local piecework or commercial farm work. This work left the women with no time to invest in their own land and they were unable to meet their basic needs. They did not receive remittances, as their children were either too young and/or had failed to leave home (see Table 7.1).

Ginny: maintaining and re-negotiating homestead access in the face of vulnerability

Ginny was about 63 in 2005 and the Ndamba *sabhuku*'s stepmother. Ginny's late husband built the house where she lived in the 1970s. She had one poorly constructed structure on her homestead. The well and toilet had since collapsed. Ginny also had arable land and a vlei garden. Ginny had a son out of wedlock in 1967. She later married into Ndamba village and bore six other children. The husband used to work as a civil servant in town while Ginny farmed in Ndamba. The husband's income funded the agricultural activities and provided money for the investment in the homestead and the children's education. Ginny's husband died in 1995. Her children stopped going to school, as she could not afford the fees. Ginny and her children survived on remittances from her son who was then married and working. In 1998, this son also died. The family's two head of cattle were slaughtered at her husband and son's funerals, respectively. Ginny had five goats and lost them to theft and illness. She has some chickens. Although her husband had been a civil servant, Ginny had not been able to claim his pension because she did not have formal proof of her marriage to the deceased man. In 2005, Ginny's eldest son's wife died. The son too became ill and lost his job in 2005. He had a rash, suffered chest pains and was coughing incessantly. He moved in with his mother so she could take care of him. The *sabhuku* initially encouraged Ginny to take her ill son to his paternal village. This was because Ginny relied on the community for assistance and the ill son drew on the meagre resources of Ndamba village. The Ndamba funeral register shows that Ginny did not contribute to the burial society. Ginny ignored the *sabhuku*. Her son's health improved briefly in May 2005 to the extent that he was herding cattle for the headman in exchange for food. He also poached fish from the commercial farms for sale. Ginny went to visit her daughter who was ill with tuberculosis and in hospital. In her absence, the ill son's health deteriorated and he died. Ginny received minimal assistance from the village residents to the extent that there were no meals for mourners. Ginny's niece from another communal area provided grain that was consumed during the funeral. Ginny did not participate in the co-operative garden. In 2003/4 she farmed only the residential land as she lacked the agricultural inputs to farm on a larger scale. She harvested one 10 kg sack of maize. She leased her arable land to another widow who compensated her with some grain. In 2004/5, although Ginny obtained seeds, fertiliser and draft agricultural inputs and draught well on time from her daughter-in law she relied entirely on her own labour and farmed very little land. The little that she cultivated was consumed while it was still green in the field[82]. The previous year she had managed to survive on World Vision handouts. In 2005, January, she was already begging for food from the other villagers. She harvested wild vegetables for consumption and worked for other villagers. Ginny received a 50 kg bag of maize at the end of August from the government as she was recognised as having limited capacity to grow her own food.

Ginny's hearth-hold experienced a downward spiral as she failed to recover from the successive shocks that shook the country. Because of the erosion of Ginny's assets, the family did not have anything to buffer them against the shocks. One of the biggest shocks to the household was the death of her husband and the subsequent loss of income. The consequence was that children ceased attending school and the homestead infrastructure crumpled. The hearth-hold was unable to invest in the Ndamba community. As a result, Ginny became increasingly marginalised and was excluded from the reciprocal systems. She could neither farm adequately nor participate in the co-operative garden as she lacked labour. As a result, she failed to take advantage of the opportunities available in Ndamba. The arrival of her ill son increased Ginny's dependence on the host community. Even though she successfully resisted attempts to evict her son from the village, Ginny was excluded from the community. Her large dependency on transfers kept her in a vulnerable state.

Rita: struggling to cope with illness and impoverishment

Rita was born in 1957. Her parents divorced when she was young. She was forty-eight years old in 2005. Rita lived alone on a *dongo* in Ndamba. She was the *sabhuku*'s stepsister. There was a single hut with a leaking roof on the property. There was no well or toilet on the homestead. She had no livestock. Rita was using the fencing poles for fuel wood. This rendered the homestead vulnerable to thieves and roaming livestock. Rita had a son out of wedlock in 1980. She sent him to live with her mother.

---

[82] Normally, to ensure food security, a household must produce enough maize. This means that there must be enough for them to consume 'green' and also have some left to dry. The dried maize becomes the household's staple until the next harvest. When a household consumes all its grain while it is still in the field, this indicates a severe food deficit.

Rita moved to the city where she secured employment as a domestic worker. In 1994, Rita married another man and moved to his village. She gave birth to three children and established an independent homestead with the assistance of her husband. Rita and her husband divorced after he married another woman. Rita returned to Ndamba, her natal village, with her three minor children and resided on a *dongo*. She was frequently too ill[83] to work and relied on the *sabhuku*'s wife and children to assist her with fetching water, cooking and care. The *sabhuku* insisted that she return her children to their father. Rita remained in Ndamba as she relied on her natal relatives to assist her in her illness. Her adult son lost his job during operation Restore Order and moved in with her. The *sabhuku* gave the son some money and advised him to go and find his own people. The son went, but failed in his bid to claim his birthright to land and returned to his mother[84]. Rita relied on -hoeing by hand, when she felt well, and used government seed and fertiliser. She harvested nothing in the 2004/5 season. She worked as a domestic worker for a teacher and for a priest in the local church. She obtained meals from the two workplaces. The work was all year during the day, with time off during weekends and evenings. This left her with little time to invest in her own homestead agricultural production. Even so, she did piecework for other villagers in exchange for food during her time off work. Ndamba village gave her access to food relief, as she was ill. Table 7.3 shows Rita's monthly (as at May 2005) portfolio. The structure of the portfolio reveals her reliance on borrowing and transfers and that, even with two jobs and subsidised meals, her income was inadequate to meet her own basic needs.

Rita's illness undermined her capacity to sustain her hearth-hold and increased her reliance on transfers. Rita's livelihood vulnerability was characterised by constant disruptions and relocations, which undermined her ability to accumulate assets. This was made increasingly difficult by the ever-deteriorating environment in which she operated. In 2005, she was ill, dependent and did not have adequate networks and financial income to keep her hearth-hold in a single homestead. As a result, Rita's hearth-hold had to be split between Ndamba and her ex-husband's home. Even then, Rita was unable to provide for herself and relied heavily on community and government aid. The community costs of paying for funerals and burial society subscriptions consumed a substantial part of her budget. The loss of her son's employment and his moving into her homestead further strained relations with the Ndamba community. Rita was continuously in debt. The nature of her off-farm activities left her with no time to grow her own food. As a result, she had to obtain food on the market. This made her vulnerable, as she did not earn enough to sustain buying food. Her case highlights the impact of illness on the hearth-hold.

---

[83] She had headaches and chest pains.

[84] Rita's son briefly secured a temporary post at the growth point but lost it after a month's work.

Table 7.3. Rita's monthly portfolio.

Income (in Z$)[1]		Expenditure (in Z$)	
Piecework	10,000	Grinding mill	5,000
Food relief		3 funeral contributions	70,000
Salary from two jobs[2]	130,000	Burial society subscriptions	20,000
Owes burial society	90 000	Groceries purchase	44,000
		Hut repair	46,000
Borrowed 2 buckets of maize for her own consumption		Hospital	45,000
Borrowed bus fare to visit her children			
Total	230,000	Total[2]	231,000

[1] Exchange rate: 1 US Dollar = 9,887 Zimbabwean Dollars at the official exchange rate and 1 US Dollar = 120,000 Zimbabwean Dollars in 2005 on the parallel market.
[2] The portfolio is subsidised by her employers who provide her meals during the day.
Source: Fieldwork 2005.

Alice: livelihood activities organised around child rearing

Alice was 36 years old in 2005. She lived in Mapfumo village, adjacent to Ndamba. Alice lived with her three children, an orphaned niece and her own brother and younger sister. The brother, a school dropout, was an active opposition party supporter[85]. The homestead was a *dongo* which Alice and her dependents had been using since the death of her own parents. Alice also spent time nursing an ill neighbour who was her major source of employment. She survived by doing piecework for fellow villagers in exchange for food. Alice had dropped out of school after the death of her father. She worked as a domestic maid in Harare. She lost this job after she fell pregnant. She left the child with her mother in Mapfumo and went to work on commercial farms surrounding Chikwaka. Alice had another child with a bus driver, who did not marry her. She also left this child with her mother. Alice's mother died in 2000, and Alice had to reside on the *dongo* to take care of her own children and siblings. In 2003 Alice married a widower in Ndamba. The man had lost his wife to tuberculosis and was ill himself[86]. The man had a well-developed homestead and thriving orchard in Ndamba. Alice moved in with him and left her other children at her natal village. The marriage produced a daughter. While resident in Ndamba, Alice was excluded from the women's garden co-operative, food relief and other activities

[85] Opposition Party (MDC) supporters were sometime denied access to government assistance (food aid, clinic, schooling).
[86] Recurrent headaches and a runny stomach.

in the village. Alice had reduced her hearth-hold's vulnerability to marginalisation within the community of Ndamba by maintaining her membership of the garden co-operative in her birth village. She also sold fruit from her husband's orchard while he was alive. Alice's husband fell ill and died in 2005. Alice's church supported her during the illness and death of her husband. The Ndamba *sabhuku* reclaimed Alice's marital home as a *dongo*. Alice returned to her parents' *dongo* in her natal village. As already stated her parents' village was adjacent to Ndamba village, and so Alice invested in cultivation, received food aid and government relief through membership of that community and was a senior member of the garden co-operative in her natal village. She had maintained her position for the duration of her marriage. Her harvest failed as she had spent most of the time when she should have been working on her crops looking after her ill husband. In August 2005, Alice's 8-year-old son fell ill. She took him to her faith healing-based church for prayers, but he did not improve. She borrowed money and took him to the clinic where they were referred to the hospital. Alice's son died before she could raise enough money to take him to the hospital. When the son died, her church, which normally assisted her with food and money, refused to participate in the funeral because she had taken the child to the clinic against their advice[87].

Alice's life was characterised by constant dislocation and disruption in response to the various calamities she experienced. Initially she tried to work as a migrant in urban areas but had to stop after she had a baby. While her mother was alive, Alice relied on her to support the children. After her mother died, Alice's livelihood options were confined to the communal area, as she had become the main caregiver for her children and siblings. The *dongo* in her parents' village enabled Alice and her children to access shelter, a community and a base from which to operate. However, with no rural-urban resource flows, the hearth-hold had limited opportunities to diversify its portfolio. Alice risked arrest by going to poach thatch from the commercial farming areas. Her initial strategy of marriage into Ndamba gave her limited access to the community benefits in the village as both Alice and her spouse remained excluded by the Ndamba community. Alice insured her hearth-hold against the exclusion in Ndamba by maintaining her position in Mapfumo, her birth village. This enabled her to gain access to government and NGO food aid. Although healthy herself, Alice's capacity to benefit from her labour was undermined by the burden of caring first for her husband and then the young children. She also received transfers from the church. The vulnerability was exacerbated by the church's withdrawal of support as a sanction against Alice as she had acted against their doctrine. Alice's portfolio reveals a hearth-hold on a rapid decline and on the way to becoming dependent.

---

[87] Some of the Apostolic Faith do not believe in conventional medicine but believe in the power of prophesy and prayer. As a result, they viewed Alice's action of taking her son to the clinic as her lack of confidence in the church.

Lizzy: maintaining a marital homestead with no income

Lizzy, a widow was more than 65 years old in 2005. She never went to school. She lived in Makuku village. She had no livestock or agricultural implements. Lizzy had the land around her homestead. Her deceased husband was an original inhabitant of Makuku. When she married, Lizzy and her husband were allocated a homestead by the then Makuku *sabhuku*. The couple lived and worked in Harare for more than 30 years. They constructed a house in Makuku using their wages to prepare for their retirement. The couple also acquired a house in Harare when the government converted the African rental housing to ownership for sitting tenants after independence in 1980. The couple had a son. When the couple retired to Makuku, their grown-up son married and took up residence in their urban house with his wife and children. The son, who was unemployed, offered no financial support to his parents. Lizzy's husband drowned in 1999. Lizzy blamed her husband's brother and his family for his death. This soured relations between the widow and her husband's family. Lizzy remained in the marital village living alone on her homestead. The homestead had a three-roomed brick and asbestos structure, a crumbling hut and a dried out well. There was no toilet. She lost all her livestock to the drought. At her age, she had no money to hire people to repair her fence and this rendered her crops vulnerable to livestock. She had a banana orchard on her homestead but never benefited from it as the fruit was eaten by cattle. When her son fell ill in 2004, she went to visit him in town. During her absence, her home was broken into and the plough and bicycle she inherited from her husband were stolen. Lizzy survived on handouts that she received from her husband's uncle who lived in Tunha village, next to Makuku. That family gave her firewood, food and money to pay the obligatory contributions in the village. They looked after her when she was ill. In exchange, she performed some light work for them. In 2004/5 she planted some old maize seed on her homestead but it was consumed by livestock. Lizzy could not remember when she last used her arable land in Makuku. The *sabhuku* sold some of her land to immigrants. She once participated in and benefited from the food-for-work programme where the most vulnerable people were selected to carry out some road repair work in exchange for grain.

The case describes the situation of a woman who because of declining health and advanced age failed to mobilise money to invest in assets to fortify her hearth-hold against successive shocks. The death of her husband coupled with the strained community relations in Makuku increased the isolation of her hearth-hold. Lizzy's vulnerability was increased by her lack of access to a reliable source of income. In their earlier years, Lizzie and her husband accumulated financial capital in the urban setting, which they invested on their rural homestead. The rural homestead was supposed to provide a place for the couple to retire to at the end of their working life. The financial flows would have continued in the form of remittances from the son and rental from the urban property. The livelihood of the homestead was exposed when financial flows ceased from urban areas as the son not only failed to secure employment but occupied the urban property which cost the hearth-hold a potential income source. As a result, Lizzy was reduced to a dependent hearth-hold Lizzy's limited capacity to mobilise agricultural inputs forced her to stop using her arable land. This rendered her arable land vulnerable to reallocation by the Makuku *sabhuku* who saw it as a *dongo*. Lizzy continued to lose her assets to the elements and livestock that roamed the village. When her son was ill, she lost her remaining moveable assets. She was entirely dependent on handouts from her deceased husband's family in the next village. Lizzy's access to the little food aid that came to Makuku was limited by the large numbers of residents competing for the food aid. The story details a hearth-hold on a downward spiral with no way of buffering itself against livelihood threats.

Tuli: displaced and highly dependent widow

Tuli was living in Makuku 3 village in a well-developed homestead. There was a brick house and a hut. Tuli was a widow whose age I estimated to be at least 80. She spoke a different dialect than the people in the study area and relied on her son to translate conversations for her. The homestead owner, Tuli's 35-year-old son, lived and worked in Harare with his family. Tuli lived on this homestead with a 31-year-old divorced and unemployed son and two teenage children who belonged to her divorced daughter[88]. Tuli moved to Makuku in 1997. Prior to that, she had her own homestead in a communal area in Mutare. Tuli's husband died during the war when her children were still at school. She managed to send her children to school through her farming ventures. As life became harder because of ESAP and recurrent drought, Tuli increasingly struggled to maintain her livelihood. In her old age, her sight and health were failing. The children had to make frequent trips to bring her to Harare for medical treatment and to maintain her homestead infrastructure. In 2003, Tuli's eldest son decided to relocate Tuli to Makuku in order to bring her nearer to his place of work where it was easier for him to take care of her. Tuli left her marital homestead as a *dongo* in Mutare. She sold her remaining livestock and relocated. In Makuku, Tuli's son purchased two acres of mostly rocky land that was not arable. They borrowed land from a neighbouring village to grow their crops (a fellow migrant). They hired draught in the 2003/4 and the 2004/5 seasons. Tuli's employed son paid for draught hire and provided seed and fertiliser. Makuku village excluded this hearth-hold from government-provided seed and inputs as they were immigrants. Tuli hired *maricho* labour. Tuli's son provided her with money to run the homestead. Tuli had a very small garden (2 x 5 metres) by the river where she grew some vegetables and sweet potatoes. The garden, like those of other immigrants in Makuku 3, was vulnerable to livestock. In April 2005, Tuli went to Harare, as her son,

[88] The daughter had since remarried and was living with her new husband.

the owner of the property in Makuku, was seriously ill in hospital with tuberculosis. In their absence, they hired labour to harvest their maize crop.

Tuli's life story is an example of how children can provide old age insurance to parents. This mediates the mother's vulnerability to old age and poverty. Tuli's son made the decision to move her to Makuku in order to bring her closer to his place of work so that he could take care of her while she in turn looked after his *dongo*. This was in response to the increasing transport costs between his place of work and the village where his mother lived. This secured Tuli's hearth-hold, as she was no longer able to meet the practical challenges of running a homestead on her own. In Makuku, however, Tuli was considered a stranger, was excluded from the Makuku community and was only able to negotiate access to arable land through the money that her employed son provided. She had no community support and lost her privileged senior woman position when she relocated. The relocation to Makuku made her vulnerable to eviction as they were settled on illegally transacted land. Tuli remained in Category III because, compared with the hearth-holds in Category II, she could no longer work for herself. However, unlike the hearth-holds in Category IV, she was receiving a regular income. The structuring of Tuli's livelihood portfolio at the time of the research illustrates the traditional ways in which rural and urban connections were managed through the flow of people and money between the rural and urban. The hearth-hold coped with deteriorating conditions by relocating the rural homestead closer to Harare and relocating his mother. Tuli's whole livelihood portfolio remained vulnerable to her employed son losing his job and/or dying. In that eventuality, she would be left with her other son and grandchildren in a hostile community and then, like the other immigrants, would probably abandon Makuku and return to her *dongo*.

Ella: diminishing returns to labour

Ella was approximately 52 years old in 2005. She lived at her homestead with her married son, his wife and child, Ella's 19-year-old daughter and four grandchildren. They lived in a single hut. They had no livestock and had sold their arable land. Ella was born in Tunha village, which is a neighbour of Makuku village. After her primary schooling, Ella married a man from Makuku. The couple constructed two thatched huts and a well on their homestead. According to Ella, her husband was never gainfully employed and never contributed any labour to the household farm. They had six children. Ella worked on farms, moulded bricks and undertook *maricho* to raise the family while the husband was not employed. In 1996, her eldest son and daughter secured employment and moved to Harare. Ella used remmittances to lay the foundation for the construction of a brick and asbestos structure. The husband died from tuberculosis in 1996. Ella has suffered from tuberculosis and chest pains since then. She tested positive for HIV/AIDS in 2003. The clinic used to treat her tuberculosis free of charge but stopped when they ran out of drugs in 2004. Ella's employed daughter succumbed to HIV/AIDS in 2004. The deceased daughter left two children in Ella's care. The son lost his job during operation Restore Order. Ella sold her arable land and invested the money in the construction of the brick and cement structure. The money was not enough to complete the building. Ella's second son married and established his own homestead in Makuku. When his wife died, that son moved back into the widow's homestead. He subsequently remarried and lived there with his wife and two-year-old daughter. Ella had another daughter living at the growth point who had two children from different men. The children lived with Ella on her homestead while their mother resided at the growth point. Ella borrowed land in her natal village to grow crops. In 2004/5 Ella relied entirely on government seed. She hand-hoed and did not apply any fertiliser to her crop. The family consumed the entire crop while it was still green in the field. In addition, she worked on commercial farms. The farmer collected them at 06.00 and returned them at 15.00 hours daily. Sometimes Ella brought brooms and mice from the commercial farms for sale. During weekends and after the harvest on the commercial farms, she performed *maricho*. Ella's hearth-hold was further reliant on handouts from the church and the widow's natal relatives between July and October when the commercial farms laid off their workers. The widow's daughter-in-law did not work on the commercial farm, as her son did not allow her. The daughter-in-law helped to look after the grandchildren when the widow was working on the commercial farms. Ella was frequently ill and missed work. Her health improved when she rested. The importance of the commercial farm income to her hearth-hold's portfolio made it imperative for her to work all the time. She went to commercial farms daily to work there. In a *kuzvarira* type marriage (see Box 5.1), Ella received bride price from a local teacher in exchange for her 17-year-old daughter. The 'son-in law' gave the widow food and provisions. However, the daughter was not interested in the marriage and attempted suicide.

Ella's life course followed a continuous downward spiral over time. Her lack of education confined her to an early marriage. The death of her husband did not have a significant impact on her livelihood, as he had never contributed to it anyway. The widow's livelihood income sources became more diversified when her children grew up and moved out of the home. Her son's loss of employment and her daughter's death halted the flow of goods and service along the hearth-hold's rural-urban connections. The vulnerability of the hearth-hold increased when the widowed son, his new wife and child and three grandchildren moved into Ella's homestead. The decision to dispose of the arable land on the market did not reduce the hearth-hold's vulnerability to poverty and income decline. This is because the money received was consumed by the increased hearth-hold population. Ella's poor health limited her capacity to exploit her labour. The collapse of the health system in Zimbabwe further undermined Ella's health. The decision to marry off her youngest daughter must be understood as a desperate act by the widow to gain access to support from someone with a regular wage income in the same way that rural livelihoods are buttressed by continued access to a reliable non-agricultural income. This was the reason the widow made such concerted efforts to make the daughter's marriage work. The decision nearly cost the daughter's life.

Betty: opportunistic trader with inadequate capital

Betty was a 36-year-old former commercial farm dweller who moved to Makuku with her husband during the 2002 land invasions. Betty lived with her 15-year-old daughter and three grandchildren. The homestead was on the stream bank[89]. They had one goat. The homestead had not been cleared of bush. There was no fence, toilet or well. There was a poorly constructed mud and thatch kitchen and a two-roomed brick and asbestos structure. Betty's parents migrated from Malawi and she was born and raised on a commercial farm. Betty had a son when she was 14 years old. She later married a long-distance truck driver and they had three children. Betty and her husband were evicted from the farm during the fast track land reform. Initially, they stayed in Makuku and later purchased their own homestead from the Makuku *sabhuku*. The couple developed their homestead with the husband's wages. Betty's husband fell ill and died shortly after, in April 2005. He was suffering from tuberculosis. Betty inherited the radio, bicycle and television set. Betty was not able to claim her husband's life insurance benefit as she did not have formal proof of her marriage. Betty was a full-time commercial farm worker. She also was suffering from tuberculosis. She frequently missed work because of her illness. Now that she no longer lived there, she made the daily trips to the farms throughout the planting season. The farmer picked them at 06.00 to return at 15.00 hours. Sometimes she stayed away for up to two weeks. Betty's 15 year old daughter also worked on commercial farms with her mother during weekends. Betty's husband's friend ploughed some of her land for her. The groundnuts and maize that Betty had planted were overrun by weeds. Her whole 2004/5 crops was a total failure Betty also engaged in the sale of seasonal wild produce, including mice, brooms and tobacco gleanings. When the bush was destroyed by fire, she borrowed some money and bought tomatoes for resale but failed to sell the tomatoes as the market was flooded. The person who had lent Betty the money confiscated Betty's chickens after Betty failed to repay the debt. When she was ill, Betty bought painkillers from the local shops. She did not have any money to seek medical attention. Makuku village excluded her from the government-provided seeds as she was an immigrant.

[89] Zimbabwean conservation legislation prohibits settlement on the river bank (Bell and Hotchkiss, 1991:204).

Betty's access to a homestead enabled her to reside in Makuku and have a base from which to operate. The death of her husband before they had consolidated their homestead left her without a reliable income and with an incomplete homestead. Betty relied on her labour, which was constrained by her ill health, to provide for her homestead. She relied on forest produce, which is seasonal, has low returns and is vulnerable to bush fires. Her attempt to diversify into tomato vending was limited by the small rural market. Betty lacked the agricultural inputs and labour needed to farm. Even though she received some assistance from her friends in the community, she failed to produce even at subsistence level. Betty invested her labour in commercial farm work. This gave her access to produce that she could sell to obtain food for her hearth-hold. Her hearth-hold's existence was vulnerable to Betty's deteriorating health. The hearth-hold's immigrant status limited their access to community assistance and government aid.

7.4.4 Hearth-holds type IV: dependent, vulnerable hearth-holds

The average age of a type IV hearth-hold head was 31 years. Type IV hearth-holds do not have an independent homestead. They exist as sub-units of other single women hearth-holds or male-headed homesteads. These hearth-holds either were living as sub-units on a homestead or were divided over different locales. For example, a single woman hearth-hold head lived in the urban areas while her children remained on the homestead in the communal areas or divorced couples where the woman returned to her parents' homestead and left the children under the former husband's care in his village. The women heading these hearth-holds were still of childbearing age. They did not invest in agriculture as they lacked land and other financial capital to secure agricultural inputs. Their assets were community membership, their kin homestead which provided them with shelter, labour and provided a fall-back position. The hearth-holds were compromised by illness or a heavy burden of care. As these hearth-holds did not control enough financial, physical and human capital to sustain their hearth-holds on an independent homestead, they coped by relying on a larger and independent homestead usually belonging to their parents. The larger homestead reduced the dependent hearth-holds 'vulnerability to the economic conditions, drought, illness and death by subsidising the dependent hearth-hold. This was through assistance with childcare, provision of shelter and food. The dependent hearth-holds' portfolios were diversified into dependence on a larger unit, petty trade activities, risky activities like prostitution, and erratic employment at the growth point or commercial farms. These hearth-holds could not sustain themselves without the continuous support of the host domestic unit. The situation was worse when the hearth-hold head was ill.

Timika: hearth-hold vainly struggling to construct a viable livelihood

Timika was 36 years old in 2005. She attended four years of school and left when her widowed mother could no longer pay the school fees. In 1983, Timika married and had two children in the adjacent village. In 1990, Timika and her husband divorced. Timika moved to her widowed mother's residence in Makuku with her children. In 1994, Timika left her children in Makuku with her mother and went to work as a housemaid in Harare, where she worked for three years. In 1997, Timika lost the job after she became pregnant and returned to her mother's homestead in Makuku. Timika secured employment as a *dongo* keeper for a person who lived in Harare. Timika then moved out of her mother's homestead and had her independent homestead on the *dongo.* Timika's employer provided her with agricultural inputs and paid her a monthly wage. The arrangement lasted for eight years. Timika augmented her income by working on commercial farms and doing *maricho* in the village. She also brewed beer with her mother. In 2002, Timika was allocated a piece of land in the former commercial farm areas by the government. She did not take up the allocation as she lacked the money and labour to establish herself there. In 2005, Timika was evicted by the Makuku *sabhuku* when he repossessed the *dongo.* She moved back to her mother's homestead. Since the demise of her first marriage, Timika had been struggling to establish a viable unit. The opportunity to take care of a *dongo* enabled her to run an independent homestead for eight years and have consistent access to non-agricultural income which is important to maintain communal area livelihood sustainability. Unfortunately, it was not enough for her to accumulate assets to establish an independent homestead as she was liable to lose it when the owners eventually retired to the communal area or as happened in this case, the *sabhuku* repossessed the land. Timika's hearth-hold remained vulnerable to the death of her elderly mother who formed the conduit for Timika's access to Makuku village. The pathway shows the risks of a divorced woman living in her mother's natal village because this makes her an immigrant and undermines her customary claim to village membership. Timika's hearth-hold survived by engaging in marginal activities that did not generate enough to sustain an independent unit. Timika attempted to reduce the vulnerability to loss of community membership by entering into the *matongo*-keeping arrangement. This arrangement enabled her to generate some non-agricultural income and diversify her portfolio. In her time as a *dongo* keeper, Timika was in Category II as she ran an independent homestead and was able to provide for her hearth-hold's daily needs. Timika slipped into Category III when the *sabhuku* repossessed the *dongo.* This cost her homestead access and an important component of her livelihood-portfolio access to non-agricultural income. The woman coped by returning to her mother's homestead which served as her fallback position. Unlike Ann, Timika was not in a position to defend her interest against the *sabhuku*'s action.

Jane: parasitic hearth-hold

Jane was 36 in 2005. She had six children aged 18, 12, 7, 5, 3 and 1 respectively. Jane had 14 years of education and was poised to marry a man of her choice. In 1987, Jane's father, who was a truck driver, had a fatal accident in which both the father and another man died. The family of the other deceased man demanded 30 head of cattle and three human beings[90] as compensation from Jane's family. In an *Ngozi* type of marriage (see Box 5.1), Jane was selected to go to the wronged family, bear children and return. Jane initially refused but eventually relented as the whole village's future was threatened by the avenging spirit of the deceased man. Jane moved to the man's village and had three children with the deceased man's brother. Two of the children survived. In 1992 after fulfilling the terms of the *ngozi* marriage, Jane returned to her widowed mother's homestead in Makuku with her children. In 1997, Jane had a daughter with a man who had been hired to herd cattle in the village. In 1999 and 2001, Jane had a relationship with another man and had two more children. She claimed she could not marry anyone because she had been spoilt by the *ngozi* marriage. During the fieldwork, she was in a four-year-old *mapoto* union (see Box 5.1) which had produced a child. The man was living with Jane and her mother at the mother's homestead. The partner could not marry her and take her to his village, as he could not afford to raise money to initiate the formalisation of the union and to look after Jane and all her children. Jane's mother and siblings felt responsible for her predicament so they assisted her to raise her children. Jane's mother looked after all her children and sent them to school and the clinic when they were not well. When the elderly mother was absent, Jane could not conduct her business, as she had to assume full responsibility for her own children. Jane made regular four-day long trips to adjacent commercial farms to trade in a variety of commodities, including produce from her mother's arable land, milk, peanut butter, drugs and beer.[91] However, she failed to generate enough income to contribute to either the host homestead and/or the care of her own children. When she went on her trips to the commercial farms, she needed someone to help her carry the goods. Sometimes she employed an orphan to assist her or withdrew her own children from school. She was arrested in December 2004 and her mother paid a fine for her.

The case shows the importance of cultural beliefs, such as *ngozi*, in lives of people in Zimbabwe. Jane moved residence to fulfil the conditions of appeasing the avenging

[90] A woman to bear at least two children.

[91] She bought nuts which she processed into peanut butter for resale. She also traded in empty grain sacks, One jar of peanut butter was traded for four empty Hessian sacks. The sacks were in turn exchanged for maize in the village at the rate of one empty 50kg bag for one (10kg) bucket of maize.

spirit. Although she was free to leave after giving birth to the children, her chances of entering into a marriage were compromised by the stigma of the *ngozi* and her children. Even though the children belonged with the deceased man's family, they followed her to her natal village and she had to provide for them. This further compromised her chances of remarriage and access to residence. Jane decided to base her hearth-hold at her mother's homestead where her natal family felt obliged to look after her and provide for her children. Jane went on to have more children in repeated attempts to remarry and secure her own homestead. When all this failed, Jane settled for a *mapoto* relationship. Burdened with six children, Jane started to make forays into the commercial farms to trade in a variety of commodities. She even took risks by trading in drugs and traditional beer. In spite of all these efforts, Jane was not able to look after her hearth-hold and relied on transfers from her mother and siblings. Jane's hearth-hold's pathway shows how livelihoods that are sorely reliant on communal area resources have little chance of establishing a viable livelihood portfolio.

Cathy: livelihood curtailed by illness

Cathy was born in 1982. She was 23 years old in 2005, a widow, and a daughter of the *sabhuku*. She spent 13 years at school. In 2000, Cathy went to work as a domestic help in Harare and lived with her employers but left this residence when she married in 2001. Cathy was married as a second wife after her husband's first wife had failed to produce children. When Cathy married this man, she went to live at the man's homestead in a communal area. The husband was working in town. The marriage produced one son. Cathy's homestead was destroyed by lightning in 2003. She moved back to town. Cathy was not working and her husband provided for her and her child. In 2003, the husband died of meningitis[92]. Cathy and her son returned to her parents' homestead in Ndamba. The widow could not access the deceased husband's pension, as her marriage was not registered. Cathy initially joined the neighbourhood police in Ndamba but left when that did not materialise into a formal recruitment into the police force. Cathy took work as a live-in domestic help in the urban area. She left her son in the care of her parents in Ndamba village. She lost her employment when the employer emigrated. She returned to Ndamba and assisted her parents with the agricultural production. She bought a goat. The goat was slaughtered at her brother's funeral. She secured another domestic employee job but lost it on May 20, 2005, after two months, when she had recurrent headaches. While at home, she was processing her mother's harvest of maize, shelling and bagging it. She secured another job, which she lost again after she suffered a minor stroke. In September, she secured employment as a clerk in Harare but, after working for only two days, she had to leave as she had a stroke. She returned to Ndamba. She brought a bucket

[92] One of the many HIV/AIDS opportunistic infections.

of wild fruit, which she went to sell at the school. She wanted to raise money so she could buy sugar in bulk and barter it for maize. In October 2005, she stopped the fruit business, as the bus fare to Harare was too high. Cathy asked her father, the Ndamba *sabhuku*, for land but he would not give her any, as he believed she might remarry. Cathy used some of her aunt's homestead to grow vegetables for sale. She borrowed some land from her friend in the next village and at the time of this research was clearing it for the 2005/6 agricultural season.

This pathway exposes a livelihood eroded by first mortality and then the illness of the hearth-hold head. The young widow's opportunities to secure her livelihood were contracting over her life course because of the death of her spouse and her possible exposure to HIV/AIDS, her responsibility for a young child and the loss of her marital home and assets. Initially it seemed she would recover and accumulate enough money to establish an independent homestead. Her constant illness limited her opportunities to diversify her livelihood, as she was constantly incapacitated. This illness cost her several employment opportunities. The widow had a relatively educated status and her networks increased her access to employment opportunities but these were all negated by her poor health status. Even when she was confined to the village, the widow tried to diversify into trading and agricultural work. Cathy's uncertain health status increased the uncertainty of her decision-making environment and her dependency on the host homestead. She lacked the basic asset of labour that other hearth-holds had in abundance. The case shows that even networks and kinship support cannot help a hearth-hold to surmount HIV/AIDS. The widow's location in Ndamba, though a useful fallback position, limited her capacity to set up an independent homestead because the *sabhuku*, her father, failed to allocate her an independent homestead. As a result, all Cathy's labour returns accumulated to the *sabhuku*'s household. Cathy repeatedly tried to secure financial capital through working, trade but her illness negated her efforts and forced her to remain in a dependent position. The case shows that networks have limited impact in reducing the vulnerability of hearth-holds that are afflicted by illness.

Netty: 'Getting by and going nowhere...'

Netty, another daughter of the *sabhuku*, was 27 years old in 2005. She was educated up to high school level. She secured a job in Harare as a housemaid. She fell pregnant and went to live with her father-in-law in Mhondoro (her husband's parents were divorced). She had a baby. She divorced that man in 1997. She brought her child to her mother at Ndamba and resumed work as a domestic maid. In 2000, Netty left to work in a shop in Mutare. In 2003, Netty moved to the growth point and secured a job as a shopkeeper through the local ZANU project co-coordinator. She had another baby. She sent the baby to her parents' homestead in Ndamba. In 2004, she left her work and secured a better-remunerated job at another shop at the growth point. The father of the young child sometimes gave her money but not regularly, as he was employed in the informal sector at the growth points. She herself had not made any investments at home, except that two years previously she had planted some maize on her father's land (got seed from the government and harvested 4 x 50 kg bags). She also helped her parents with expenses for thatching, plastering and glazing their homestead. She was renting a place at the growth point where she was working. She contributed regularly to the Ndamba village burial society.

Unlike her sister Cathy, Netty was not ill, so she was able to work for longer periods. However, after ten years of working she had not accumulated assets to either establish her own homestead or make regular contributions towards the upkeep of her children in Ndamba. She made erratic investments in her parents' homestead and contributed to the Ndamba burial society to ensure her community membership. In spite of her several networks that facilitated her access to employment opportunities, and relatively better education Netty's hearth-hold remained as dependent as those other hearth-holds in this category. This shows that for the younger hearth-holds in the sample, employment did little to mediate vulnerability to the declining economy and inflationary pressures.

Holly: 'divorcee' starting out

Holly was 19 years old in 2004. She dropped out of school as her brother who was paying her school fees lost his job during operation Restore Order. Holly went to live with her widowed mother in Makuku. The mother, Holly, Holly's sister's four children, a married brother and his wife all occupied a single hut on the homestead. The widowed mother married Holly off to a local teacher in a *kuzvarira* type of marriage (see Box 5.1). Holly was not interested in this man and refused to move in with him. The mother tried to force her to go. Instead, Holly went to Harare and secured employment, but the mother followed her and brought her back, as so she could live with her husband. Holly secured employment as a convenience shop [93] assistant in Makuku but the shop was closed during operation Restore Order, so Holly lost the job. Holly's mother kept pressurising Holly to move in with 'her husband' as the man was now supplying the homestead with food. In September 2005, Holly attempted suicide. A village court was held and Holly's marriage was annulled. Holly left Makuku and went to live with her elder sister at the growth point, where she secured employment in a shop that sold beer. Holly earned the equivalent of five United States dollars a month. She lost that job after a month as the employer also expected her to 'entertain'[94] the clients. She returned to Makuku village and went to live at her mother's homestead where she assisted her mother with the agricultural tasks.

---

[93] A shop set up at a spot that has not been designated for trading by the spatial land use planners. In this case, it was in a residential zone and was therefore an illegal development.

[94] The employees are expected to flirt with clients and humour their advances so as to attract more clientele.

This story shows the extent to which hearth-holds vulnerable to poverty will go to sustain their livelihoods. Holly's life story as a single woman, though still in its infancy, illustrates the extent to which women can be controlled not only by men but also by other women. This limits their decision-making space. The consequences are that these women's decision-making space is curtailed, leaving them at the mercy of those who wield power over them. In just a period of a single year, Holly's status changed from schoolgirl to married and, eventually, divorcee. Although she did not actively set out on this path, she had to deal with the consequences, with the result that she attempted to take her own life. Holly's attempts to get on with her own life and construct a livelihood were curtailed by the mother's interference, operation Restore Order and poor working conditions. The result was that she ended the year 2005 in the same position as she started: a wholly dependent person.

Tabitha: lady of the growth point

Tabitha stopped going to school after her father died. She was 38 years old in 2005. In 1985, she left Makuku and went to work as a domestic worker. In 1991, she became pregnant and moved back with her mother. After the birth of her daughter, she left the child with her mother in the village and went to work on the commercial farms. In 1998, she met a man and entered into a *mapoto* marriage (see Box 5.1). The marriage produced a son. They moved into rented accommodation at the growth point where they were trading in beverages. Tabitha's partner fell ill and he died from tuberculosis in 2000. She took her son to her village and left him with her widowed mother who lived in Makuku. Tabitha secured employment in a liquor shop at the growth point. She left that job in 2003, as the pay was too low. She set up her own shop, cooking and selling food to passing truckers. She stopped selling food during operation Restore Order. In August 2005, she started going to the commercial farms, buying cabbages for resale at the growth point. She was renting a room at the growth point, which she shared with six other women. She had never engaged in agricultural production but used to buy agricultural inputs for her mother when they were available. She visited her children twice in six months. She also relied on transfers from the communal area for staple maize.

The pathway reveals a livelihood that is struggling with vulnerability to impoverishment, death and the harsh economic conditions prevailing in Zimbabwe. Tabitha left the village and constantly struggled to stabilise a livelihood that remained largely instable and insecure. In spite of this, Tabitha was determined to continue residing at the growth point and exploiting the livelihood opportunities there. The rate at which she changed her livelihood portfolio highlights the rapid decline in the economy and the increasingly difficult operating environment. Operation Restore Order disrupted a relatively high input-high return business that she had established. Thereafter, she had to start experimenting again. As a result, her portfolio was made up of short-term, low-return activities. Although she lacked any short and long-term assets, her mother's homestead in Makuku remained a stable fallback position to which she resorted in times of crisis and relied on for the upkeep of her children. Tabitha, like Netty who was also based at the growth point, could not meet the daily requirements of her hearth-hold and remained heavily dependent on the family in Chikwaka communal area.

Asher: matongo *employee*

Asher was 35 years old in 2005. She was living alone on a homestead in Makuku 3. She was employed by the owner of the *dongo* who lived and worked in Harare. Besides keeping the *dongo*, Asher sold second-hand clothing on behalf of her employer. Asher became a widow in 2000 when her son was six years old. In 2002, Asher left her son under the care of a grandmother and moved to Makuku 3. In 2003, Asher married her employer's brother. The husband, who worked as a cobbler at the growth point, was a widower. The couple did not have children. Asher's husband was coughing and had chest pains at the end of 2004. He relocated to Harare to seek medical care. Asher's husband died in February 2005. Because of her husband's illness, she was not able to do her work. She spent all her time visiting him in Harare. She tried to do some manual ploughing and planted seed from her granary because she missed the seed distribution while she was visiting her ill husband. The maize crop was late and had not been weeded. When I visited, Asher and her deceased husband's sister were packing the late man's belongings so they could ferry them to Harare for distribution. Asher was hopeful that she would be allowed to continue to stay here and keep the land for her sister-in-law, who was also a widow.

Asher's pathway details an insecure livelihood that has failed to stabilise despite her securing employment. This is a common characteristic of all the single women whose main livelihood activity is employment. Asher's decision to remarry culminated in her becoming a widow and being exposed to HIV/AIDS. The changed marital status also changed her relationship with her employer and made her terms of employment uncertain.

7.4.5 Hearth-holds type V: succumbed

Chapter 2 reported that the research sample decreased as hearth-holds from the sample succumbed to the difficult conditions. The hearth-holds that succumbed to the conditions represented the worst-case scenario for single women. At this point, a hearth-hold has no assets left to protect itself against threats to its livelihood. The following case is presented to identify the circumstances that can overwhelm hearth-holds.

Tina: hearth-hold decimated by illness

When I met Tina in 2004, she was 32, ill and bed-ridden. She lived with her nine-year-old daughter on her parents' homestead in Makuku. Tina s parents employed a divorced woman to look after her. Tina was born into a well-to-do family and received a good education. She married when she was 20 and had two children. She secured employment as a civil servant, lived and worked in Harare. Tina and her husband divorced. Tina got custody of the children and moved into her parents' house in Harare from where she continued to work. In 2003, Tina was diagnosed with pneumonia. Her then six-year-old daughter continued to lose weight and was generally unwell. She had a poor appetite, was weak and small for her age. In 2004, too ill to work, Tina and her daughter moved to her parents' homestead in Makuku. Tina's other siblings paid for all her medical expenses and assisted her parents to look after her. Tina and her daughter had access to medical care in Harare. In June 2005, the condition of Tina's daughter deteriorated and she was rushed to hospital where she died. She was buried in Makuku. The daughter's father could not attend the funeral, as he was too ill. Tina slipped into a coma shortly after the burial of her daughter. The family transferred her to hospital in Harare. She died in August 2005.

The case highlights the consequences for homesteads that are vulnerable to HIV/AIDS. Tina's hearth-hold first lost its own individual assets and then had to rely on assistance like the women in category IV. However, even with the extensive family support and access to medicine, the hearth-hold was unable to cope and finally succumbed to HIV/AIDS. Tina's case brings to the fore an emergent consequence of the virus, the illness and death of very young children in addition to the economically productive people.

7.5 The influence of children on hearth-holds

The circumstances of children also influenced single women's livelihood decision-making. Children present women with both opportunities and challenges. Single women hearth-holds' children are a source of insecurity and vulnerability. On the other hand, resident children provide companionship and much needed labour but at the same time increase a pressure on a hearth-hold's livelihood portfolio. Unfortunately, children also pose serious mobility constraints to their primary caregivers whose activities are limited to the locality. The indeterminate stay by adult children in the hearth-holds' homestead increases the uncertainty of the women's decision-making environment. In the current study, the hearth-holds with dependent adult and young children are in Categories III-V. The hearth-holds in Category IV rely heavily on transfers. At the time of this study, they were all still experimenting with opportunistic activities such as trading and shop keeping which did not bring in an adequate income for them to establish and maintain homesteads for their hearth-holds.

7.6 Discussion and conclusion

The chapter set out to establish how single women hearth-holds make livelihood decisions under the combined threats of HIV/AIDS and worsening livelihood conditions in the communal areas of Zimbabwe. Using the hearth-hold as a unit of analysis it was shown that under the existing volatile conditions in Zimbabwe, hearth-hold vulnerability cannot be exclusively attributed to HIV/AIDS. However, HIV/AIDS worsens the plight of already vulnerable people by increasing mortality, illness, burden of care, displacement and the erosion of human capital. Single women have to make decisions and construct livelihoods in an increasingly volatile environment where HIV/AIDs and worsening livelihood conditions increase pressure on hearth-holds livelihoods. The pathway studies of 22 women illustrate the different ways in which hearth-holds made decisions to cope with the volatile situation. The extent to which hearth-holds could diversify their livelihoods to mediate vulnerability to the volatile conditions depend on their health of members, the hearth-hold's assets, age of the head, access to non-agricultural income, local governance and the specific shock to which the hearth-hold was exposed. A typology of hearth-holds shows that the hearth-holds which made the decisions that mediated vulnerability, maintain and sustain their livelihoods were category I while those that succumbed to the pressures on the livelihoods were in category V. The typology shows that the younger hearth-holds were generally more vulnerable and were less equipped to mediate vulnerability to shocks over the life course. This underscores the importance of experience in hearth-hold livelihood decision-making. However, the finding that Ginny a sixty three year old hearth-hold head is in category III means that no single factor is the most important determinant of hearth-hold livelihood decision-making.

Access to a community and residence in the village gave the hearth-holds a base and enabled women to access support, networks and other livelihood opportunities. The hearth-holds that managed to negotiate access to an independent homestead were able to make autonomous decisions. The hearth-holds' pathways were useful in highlighting the importance of the communal area homesteads not only as a primary residence for the hearth-holds but also as an important fall-back for hearth-holds who respond to the volatile conditions by relocating. The communal area residence served as a repository for hearth-holds to leave their young children while the hearth-hold heads went in search of livelihood opportunities. The capacity to negotiate access to an independent homestead depended on a hearth-hold's age, timing of shocks access to non-agricultural income and the specific local governance conditions of the village.

Although communal areas are governed by the same formal legislation, the specific local context of the villages presents opportunities and constraints for the hearth-holds. The women in Ndamba reduced vulnerability through a cohesive community, networks and increased access to initiatives of non-governmental organisations.

These initiatives provided single women hearth-holds with alternatives that mediated their vulnerability to agricultural cop failure and drought. The small population in the village increased the opportunities for hearth-holds in Categories III and IV to benefit from any government handout. This was because the small number of homesteads could all have a share of anything even if it was a single sack of maize seed. In Ndamba, the small community enabled hearth-hold to receive some form of obligatory kinship help to cope with shocks. The absence of a land market in Ndamba, although increasing people's access to thatch, fuel wood and grazing.

However, the Ndamba governance regime also undermined hearth-holds' livelihoods and increased d their vulnerability to the volatile conditions and HIV/AIDS. The absence of a land market denied some women in Category III opportunities to dispose of their idle arable land in exchange for cash. They would then have been able to use the cash to pay for an emergency, such as a funeral, or to diversify their portfolio. By restricting single women's autonomy in decision making, the Ndamba community undermined single women's ability to cope with shocks brought on by drought, death in the house or the loss of an employment opportunity. This included restricting the number of dependents hearth-holds could host, forbidding land sales and discriminately applying the regulations that govern access to village membership, residence, funeral society membership and reciprocal networks. This led to the marginalisation of people like Alice and hearth-holds' illegitimate children.

In Makuku village, the *sabhuku* allowed an illegal land market to flourish and did not control the population in the village. The relatively high number of people in Makuku village resulted in higher pressure on land and increased competition for the scarce government handouts. The Makuku land market depleted natural capital but brought other opportunities for employment on *matongo.* The land market enabled some women to generate income by leasing out their land and disposing of land on the market. The opportunities to participate in the land market enabled some single women to reduce vulnerability by disposing of and leasing out land. This was beneficial for single women hearth-holds in Category I.

The larger population increased tensions over the selection of beneficiaries for the distribution of food aid (see Chapter 4). In order to cope with the competition, the Makuku community excluded immigrants from government aid. This increased the vulnerability of immigrant hearth-holds as this source of transfer made little impact in reducing the vulnerability of hearth-holds in Category III. The land market was a source of risk for women who lost their homestead access on the basis of *matongo* because of the *sabhuku*'s disposing of land. The hearth-holds in Category III who disposed of land in Makuku lost this valuable asset and increased their hearth-hold's vulnerability to landlessness. Although all land in Makuku was vulnerable to the *sabhuku*'s illegal land-selling activities, senior women who were related to the *sabhuku* (his mother and aunt, both in Category I) were able to resist his attempts to dispossess them of their land. Other hearth-holds were not as fortunate as their idle arable land was disposed of by the *sabhuku*.

Government failure to provide health services has increased the vulnerability of the populations. The harsh economic conditions which have increased the loss of employment by able bodied people and HIV/AIDS prevalence have changed the dynamic of gendered reproductive responsibilities. Instead of women being looked after by their grown-up children, it has become increasingly normal to find adult children depending on their mother. This phenomenon, besides extending a woman's burden of care and reducing single women's access to remittances and household labour, undermines single women's capacity to diversify their livelihood portfolio. The cases reveal that women who were looking after ill adult children and/or young children were restricted in their opportunities to diversify because they had to provide care.

High HIV/AIDs induced, mortality, marital dissolution, and ill health decimate the human capital. The displacement of large volumes of people, the illegal land market, and dwindling flow of goods on the rural-urban nexus government regulation of trading, the poor rural markets, and the hyperinflationary environment all fuel communal area livelihood insecurity.

The hearth-holds' pathways showed how diversification from risky activities enables hearth-holds to mediate risks. Generally diversification involved a shift from agricultural activities. This was because agricultural activities were undermined by the frequent drought and the declining capacity of hearth-holds to secure agricultural inputs on time. This increased the hearth-holds vulnerability to crop failure and starvation. Single women in the study were forced by circumstances to participate in the economy outside the communal areas on adverse terms as they lacked the skills and experience necessary to navigate this environment. The government legislation on trading imposed further constraints on single women's ability to explore opportunities beyond the increasingly vulnerable communal area. The hearth-holds took risks and traded without licences. The women risked arrest. When unlicensed traders' businesses were shut down during operation Restore Order, some hearth-holds which had been employed by these lost their employment.

The pathways of the successful hearth-holds show how they successfully diversified their livelihood portfolios over time. This diversification has allowed them to accumulate movable and immovable assets that they can liquidate to mediate the hearth-hold's vulnerability to loss of employment, an adult death or a drought. As a result, women in Category I not only maintained but in some cases consolidated their portfolios after shocks such as the loss of a spouse, relocation, loss of an adult child and the hyperinflationary environment. Hearth-holds in Categories II-V are on pathways that have deteriorated over their life courses. Although they differed in their experiences, these women had become increasingly vulnerable over the years to loss of assets, including loss of a spouse, HIV/AIDS, loss of livestock due to drought/theft or natural causes, disposal of land on the market and the end of marriage. The single women were unable to mobilise and/or maintain livelihood capitals to secure the livelihoods of their hearth-holds. They increasingly relied on transfers from the State, church or other households. They had a history of working on commercial farms, multiple failed marriages or their livelihoods had declined over time as they had to deal with sustained human capital loss through mortality and morbidity. The women had never had much capital in the form of supportive families or other networks. The women lacked access to reliable non-agricultural income because their children were too young, or because the single women had lost access to non-agricultural income through the death or retrenchment of adult children.

The portfolios of all hearth-holds show that, although an increasingly risky activity, agriculture remains an important source of subsistence. The successfully diversified hearth-holds were also the best agricultural producers. They had access to the resources necessary to farm. Such hearth-holds all produced adequate harvests for their subsistence, and sometimes surplus. These hearth-holds continued to farm because, for them, it was a less risky way of providing food than trying to source it through an increasingly volatile market. The hearth-holds in Category II diversified from the high risk associated with agricultural production by engaging in activities that were not directly exposed to agriculture. These included working for those who had adequate resources and getting payment in produce, and producing and bartering commodities such as pottery, sacks and groceries in exchange for agricultural produce. In this way, hearth-holds ensured access to subsistence food.

Hearth-hold livelihood vulnerability arise from the volatile economy, HIV/AIDS and single women hearth-holds' increasing burden of care and the changing nature of rural-urban connections. Through the 22 pathways, it has been possible to examine the role of networks in mediating livelihood vulnerability. Networks helped to mediate vulnerability by increasing single women's access to information and land for a garden in Ndamba, and by providing opportunities to diversify their portfolio and coping mechanisms. It was through networks that women obtained opportunities to work at the growth point, the school, urban areas, engage in cross-border trading, and diversify into working for the spirit medium and as providers of special ritualistic services. Although networks go a long way in mediating vulnerability, there are limits. In Makuku, a diversified community excluded immigrants from their community benefits, and Ndamba limited the beneficiaries by controlling access to the village. The relatively healthy single women were the ones who were best placed to exploit opportunities identified through networks. The tasks that the single women performed were physically demanding and this increased the vulnerability of single women who were ill or aged. Networks offered little in the way of assisting single women who were ill and/or had contracted HIV/AIDS.

Though some hearth-holds in the sample successfully consolidated their livelihoods through diversification, this alone was not adequate to ensure livelihood viability. Hearth-holds starting from a poor asset base have limited opportunities for diversification. The experiences of single women hearth-holds who tried to diversify into multi-spatial livelihoods failed to protect women from the livelihood threats. The rural-urban connections enable the inquiry into the whole spatial dimension of livelihood organisation in Zimbabwe. Through the hearth-holds pathways, it was possible to establish the ways in which these women maintained the rural-urban connections outside the marital relationship. In a marriage relationship, the roles of earning a wage and providing shelter lay with the men in urban areas while the women remained in the rural areas to maintain the homestead, farm and raise children. Over time the unit exported children to the urban areas to be educated and eventually the parents retired to the rural home where they survived on remittances sent by the children. The situation in Zimbabwe that virtually ended the cyclical migrant labour contract undermined all these linkages. Through the pathways of hearth-holds it was possible to see that formerly married women whose husband had earned a wage had been able to accumulate assets in the traditional way described above. However, for the homesteads that had remained exclusively in the rural areas, it had not been possible for them to develop without exploiting a market niche that secured them access to goods and services on the rural-urban nexus. Women on their own could not necessarily use the same strategies. As a result, the nature of single women hearth-holds' rural-urban connections range from those that managed to earn a reliable remittance and or rental income to those who relied on opportunistic and erratic flow of goods and services. The latter hearth-holds struggled to maintain a rural homestead and have secure livelihoods. This shows the interdependent relationship that exists between the rural and urban and confirms that one cannot exist without the other.

Vulnerability has many dimensions. Age for example, reduces vulnerability on the one hand but also increases a hearth-hold's vulnerability on the other. Shona women gain seniority with age. Senior women (in age) in this study were able to accumulate control of land through a combination of upward inheritance and assuming control of *matongo*. The women also used their senior position to defend their land interests against the *sabhuku* and other men in the community. The postmenopausal women were hired to brew beer and perform rites for the dead. In return, they were compensated with livestock or other commodities, which added to their hearth-hold portfolios. Seniority presents opportunities for single women to resist processes that might threaten their livelihoods, such as the disposal of land by the village head in Makuku and attempts to deny single women opportunities to care for their unemployed or ill children in Ndamba. Senior women are also vulnerable to accusations of witchcraft, which weakened their negotiating positions and resulted in some being ostracised.

The experiences of the 22 hearth-holds reviewed here show that, although people use various livelihood capitals to construct and maintain livelihoods, there is no single resource that is adequate to ensure livelihood success. The women in this study relied on a variety of resources in diverse portfolios to maintain their livelihoods. The chapter identified, through single women's pathways, the ways in which the hearth-holds used their resources to confront their livelihood risks though not always successfully.

Chapter 8
Discussion and conclusion

This thesis has focused on the ways in which single women take decisions to cope with risk and gain access to resources in communal farming areas of Zimbabwe. A study into the experiences of single women hearth-holds was necessary because there was a dearth of systematic research on how these women accessed resources in a patriarchal and legal space. The general situation with regard to communal area livelihood decision making has a bearing on the situation of single women. Communal areas have been undergoing significant changes as a result of government policies, HIV/AIDS, erratic weather, ambiguous governance, the collapse of the agricultural economy, hyperinflation and the general deterioration of the Zimbabwean economy over the last decade. The changes which include structural adjustment policies, land reform, recurrent drought and HIV/AIDS undermined the structural arrangements that enable people to accumulate livelihood capitals. One of these was alteration of rural-urban connections which underpinned both rural and urban livelihoods in Zimbabwe. Since the rural and urban are a single social universe, the collapse of one would inevitably undermine livelihoods in the other as flows of goods, financial capital and labour are disrupted. Another consequence was the return to the communal areas by men who failed to secure waged employment in the urban areas. This combined with the high death rate, an uncertain decision-making environment seriously undermined communal area livelihoods. The inhabitants responded by diversifying their activities and by renegotiating the flow of goods through rural-urban connections. The location of the research area in close proximity to a commercial farming area, a growth point and Harare the capital city increases the community's potential for diversification of livelihoods. The scope for diversification locally is limited by the agro-centric nature of activities, strict legislation and the local market's dependence on agricultural income.

A secure access to land is important for community membership and residence in the community. This enables hearth-holds to have residence, agricultural land, networks and a place to fall back on. This study has established that, although communal area resource governance systems are male biased, within those systems, there are more diverse forms of access to land than previously thought, and single women have more room to manoeuvre than previously thought. The norms used to determine single women's access to resources are flexible and negotiable. However, the flexibility of the communal area resource governance has simultaneously increased both the opportunities and the vulnerabilities of single women's access to residence in the communal areas. This contradiction has arisen because of the peculiar dynamic of different processes that influence resource access in these areas and the specific circumstances of single women at any one the time. An example is how the death of an adult male increases his widow's risk of abandoning the residence, while at the same time such abandoned residences potentially provide land access for other single women displaced by divorce and/or illness. However, it is not possible to make generalisations about the vulnerable people's response because the difficult conditions have altered livelihood strategies.

8.1 Resource access of single women in communal farming areas

Generally, single women have access to land in communal areas. This study has shown that single women gain access to residence in the communal area through kinship, the market and opportunities that arise as a result of the high mobility and mortality of communal area inhabitants. These opportunities, discussed in Chapter 6, are known as *matongo*. Single women's birthright entitles them to land in their natal village, in the form of male kin's obligation to look after their female kin as daughters and siblings in their time of need. This claim co-exists with patrilineal transfers of land through inheritance to ensure that single women have a means of survival and a place to reside in their natal communities. The women's rights remain hidden in investigations that emphasise the use of formal resource access investigation frameworks to analyse land access. The formal frameworks dismiss or deny the importance of these so-called secondary rights. Women's land claims are not handled in the same way as men's, but can be enforced socially and customarily. These claims, though weaker than men's, are significant as single women cannot be thrown off the land, and the women can use these claims as a negotiating base for access to other resources, such as assistance with children, illness and livelihood organisation. The finding that widows who remained in their marital residence after the death of their spouse constitute the majority of the single women in this study highlights the importance of the marital relationship for women's access to residence. The finding that divorced women were generally accommodated among their birth kin emphasises the importance of the natal relationship for divorced single women's land access.

Even though the kinship-based modes of land access remain important for single women, new and innovative ways of accessing land have been introduced through market transactions, employment relations and rights of dependents of single women. The market and employment are the only means of access available for single women who cannot enter communal areas on the basis of marriage or kin relations. Market access has become increasingly important for former commercial farm workers and immigrants from other communal areas of Zimbabwe. However, the land market has increased poverty levels and vulnerability as land that has been sold is no longer available as a fallback for single women. There appeared to be a trade-off between the market and social security benefits for single women in the study villages. The market provided residents with opportunities to raise cash through land sales and employment, but this came at the expense of the social security which was eroded as communities became more heterogeneous. This has undermined kinship ties and social security systems which work through the obligations and entitlements entailed in these systems.

The studied single women hearth-holds' residence access in the communal areas was contingent upon sets of factors that included marital status, age, child-bearing status, fertility status, rank and access to non-agricultural income. The constantly shifting conditions induced by mortality, rural–urban connections and opportunistic decision making by those in power has increased the instability of single women hearth-holds' resource access in communal areas. The different ways in which single women access land, from allocation of primary rights to the possibility of divorced women maintaining their marital residence, reveal that anything is possible because existing rules are manipulated, contradictory and used strategically by, for example, the village heads to gain patronage and or control of land resources for their personal gain.. In this study, single women's vulnerability either increased or was mediated depending on their relationship to those in power. However, the conclusions should not be overly optimistic. Even though kinship now provides relatively secure land access for single women, this security is contingent. Experience suggests that if there is a contestation between males and females, the female claims will be subordinated to the males. The rapidly changing conditions in communal areas simultaneously undermine existing conditions and give rise to new ones and new patterns of vulnerability.

Gendered resource access organisation in communal areas has been shifting because of increased poverty. This has caused the evolution of the production systems based on agriculture, and changes in patterns of migrant labour. The end of the migrant labour contract precipitated by the collapse of the economy resulted in limited cash flowing to rural areas for investment in agricultural activities. As a result communal area land ceased to function primarily as an agricultural asset. This has produced an enormous diversity in the composition of households, hearth-holds and homesteads, life cycle dynamics, and the meaning of land in communal areas. For many years, the structure of agricultural systems maintained and perpetuated male dominance and the subordination of women in resource access in communal areas. Now, production units are no longer dominated by marital unions but rather by a diversity of arrangements that include single women hearth-holds, skip-generation households, individuals living alone and child-headed households. This happens because the males are often absent from the communal areas, deceased, or have become dependent on the women hearth-holds who have to assume an increasing burden of care. This diversity of production unit means reduced control by males in the immediate unit, although males continue to dominate the wider institutions governing resource access, such as villages and family.

Studies of women's land-right vulnerability in the developing world overemphasise the role of patriarchy. The historical evolution of the communal areas illustrates that patriarchy is only one of several institutions governing resource access in Zimbabwe's communal areas. These areas are characterised by a variety of governance structures and institutions which draw power from a variety of sources, including the government, the dominant political party, traditional authorities and formal legislation. The structures and systems have multiplied over time. The diversity of authorities, lack of clear and systematic rules, the decentralisation of the decision-making process from the wider community to individual families and the diversity introduced by in-migration and marketing of communal area homesteads have rendered legal systems and rules largely irrelevant in determining resource access in communal areas. This has provided opportunities for people to negotiate resources through other institutions and arrangements. The scarcity of cash and the increased demand for land provides space for opportunistic behaviour by those in authority. It is not possible to generalise on either the processes or the outcome. The constantly shifting conditions induced by mortality, migration and opportunistic decision making by those in power have increased the instability of single women hearth-hold's resource access in communal areas.

The study of two different villages, Ndamba and Makuku, located in the same ward demonstrates how different modes of local governance influenced resource access and increased the diversity of outcomes for single women in the same communal area. The perception and interpretation of the laws and legal institutions by those in authority, and the ability of the single women to negotiate the range of institutions and authorities also increased the diversity. The Ndamba *sabhuku* suppressed land market activities by discouraging land sales and transfers and by imposing strict control on the movement of people into the village. As a result, the village remained relatively small, and more cohesive. They maintained a close-knit community which for now appears able to deliver more social security benefits to its residents than Makuku village. The exclusionary practices of the Ndamba village head made single women's children more vulnerable and undermined hearth-holds' stability in ways that could lead to displacement of single women. Makuku village on the other hand tolerated the land market, and as a result had a larger and more diverse population than Ndamba. The Makuku example shows how the land market increased the diversity of ways in which people accessed *matongo* but also increased conflict between the different people. This opened opportunities for outsiders to access land. Single women hearth-holds are among the outsiders who access land on the illegal market. The Ndamba style of governance shows how *matongo* can be managed to provide increased social security for single women not only on the homesteads, but also on the fallowed land by promoting access to natural resources. The absence of a land market in Ndamba increased social security for those having kin relations with the inhabitants of Ndamba, as *matongo* were retained in the family. This meant that such homesteads continued to be available as a fallback for vulnerable single women belonging to the kin group. The fallow agricultural land which reverted to bush provided wood fuel and thatch for Ndamba residents. The available land and small and cohesive village enabled the village to take advantage of various initiatives offered by non-governmental organisations (NGOs). However, this limited women's access to residential land, which was important for negotiating access to other resources.

8.2 Livelihood decision making of single women

The livelihood decision making of single women examined through their life stories shows how single women managed in the fast-changing environment. The study found that, although the importance of land as an agricultural asset had declined for single women in the communal areas, access to residential land and the community was a basis for negotiating access to social security and livelihood opportunities. However, this in itself did not guarantee secure livelihoods.

The single women from the two study villages attempted to secure livelihoods in hostile conditions characterised by an eroding resource base and increased burden of care. The resource bases were undermined by the failure of the domestic units to export labour which has historically been a source of remittances. HIV/AIDS undermined human capital through illness and death. The contraction of wage-earning opportunities, caused by the economic collapse of the country and government-led land reform, further limited the communal area residents' access to diversification opportunities. The increased competition for the local market undermined the viability of communal area-based retail activities. The continued use of laws to suppress trading in communal farming areas further undermined single women's opportunities for diversification. All the foregoing, coupled with the opportunistic behaviour of the local leadership, has eroded the resource base of communal areas. Subsistence agriculture – a source of basic food – was constrained by limited access to agricultural inputs, successive droughts and the decimation of labour by HIV/AIDS. This significantly reduced resources locally available for people to insure themselves against risk and limited the scope for livelihood diversification. The failure by the State to provide healthcare increased the women's burden of care. Both young and adult children relied on the single women, and this influenced single women's decision making in various ways. The research has shown that increasing unemployment and HIV/AIDS increased the dependence of adult children on their mothers. As a result, women past their child-bearing years were assuming responsibility for the care of their children and grand children.

The pathways of the single women in this study illustrate the limited opportunities for diversification available to the increasingly vulnerable women. Each of the 22 cases was unique. The extent to which single women could diversify their livelihoods depended on the hearth-hold's assets, health of members, developmental stage of hearth-holds, access to non-agricultural income, access to residence, local governance and the specific shocks to which the hearth-hold was exposed. The continued exposure of livelihoods to shocks further depleted the resource base and limited the diversification options. The portfolios of single women reflected the reduced importance of subsistence agriculture because of the increasing risk associated with agriculture and single women hearth-holds' limited capacity to mobilise the labour and agricultural inputs needed to engage in agriculture. The affected women resorted to other means to secure basic food, including clustering around those who could produce, relying on transfers from the State and engaging in illegal activities. The better-off hearth-holds had access to non-agricultural income, were elderly and generally in good health. The single women who were not coping were comprised of old and younger women who were either ill themselves or were caring for ill and/ or young children. These desperate single women were particularly vulnerable as they entered into risky enterprises such as transactional sex, which increased their vulnerability to bearing more children that are dependent and to contracting HIV/ AIDS. The difficult circumstances of these single women resulted in some of them undertaking risky activities like illegal trade in beer and commodities, withdrawing children from school, prostitution, poaching fuel and thatch from commercial farms and working when ill. These activities were not sustainable and increased the vulnerability of hearth-holds. The local governance conditions in Makuku and Ndamba villages increased and curtailed opportunities for diversification by encouraging and repressing the land market, respectively. Although the land market in Makuku provided diversification opportunities, it increased competition for the scarce government transfers and undermined the community's capacity to engage in co-operative activities. The absence of a land market in Ndamba denied women the option of disposing of their land to cope with a calamity. The relatively homogeneous community of Ndamba reduced vulnerability by providing networks through which members received transfers and access to resources. The community was also better organised to benefit from transfers from the State and limited NGO initiatives. The limited and increasingly risky livelihood diversification environment in communal areas forced some women to explore opportunities outside the communal areas by pursuing multi-spatial livelihoods. In order to cope with the limits imposed by the communal area environment and to reduce risk, hearth-holds spread their activities over different locations, extending to growth points, former commercial farming areas, urban areas and neighbouring countries.

In order to cope with the limits imposed by the communal area environment and to reduce risk, hearth-holds spread their activities over different locations, extending to growth points, former commercial farming areas, urban areas and neighbouring countries. The concept of rural-urban connections was used to capture the diversity of movements of goods, people and services between communal areas, commercial farms and the growth points. Historically the communal areas were established and designed to develop on the basis of these rural-urban connections. While the exact nature of the connections varied between individuals, the general pattern was that communal and urban areas were linked economic spaces which were interdependent. The various activities and exchanges between the communal areas and other spaces in the country enabled the people to sustain livelihoods in communal areas. This included income, people, goods and services. The decline of the urban wage labour economy and the increasing number of single women changed the nature of rural-urban connections. Historically, the men moved to earn a wage while women remained in the rural areas to sustain the homestead. The research established the importance of urban connections for the communal area livelihoods. The increasing insecurity of urban livelihoods was mirrored by the deterioration in rural area livelihoods. Unlike the men who in the past left a spouse in the rural areas to raise the children and maintain the homestead, single women hearth-holds' rural-urban connections were configured differently. This was to enable the women to meet their reproductive obligations in communal areas and secure services and goods from the urban areas to ensure the viability of their livelihoods. Single women who had a reliable access to remittances and/or rental form an urban property managed to maintain and extend their homesteads and livelihoods diversified into accumulation. The other strategy that hearth-holds in communal areas used to secure their livelihoods was to indirectly access money through working for those who had direct flows from urban areas or through the institution of *matongo*. Generally, the hearth-holds with erratic flow of goods and services on the rural-urban continuum had the least viable livelihoods. Single women tried to secure the flow of goods and services by moving out, splitting hearth-holds and engaging in trade. In addition to the single women hearth-holds, some male headed domestic units relocated to urban areas. This increased the incidence of *matongo* which enabled single women to reduce their vulnerability in the communal areas.

The different sources of income included receipt of rent from urban properties, waged work at growth points, commercial farm work and retail activities on the farms and at growth points. The single women's capacity to explore multi-spatial livelihoods was limited by the presence of dependent children, the women's age, the health of the hearth-hold's members and the women's limited labour skills. The returns from these multi-spatial livelihoods were inadequate for the women to consolidate their livelihoods. The women who were wholly dependent on mobile activities had some of the most vulnerable livelihoods. Some of them failed to secure independent homesteads in the village and remained sub-units of other homesteads. Generally, the combination of localised and distant livelihoods failed to protect women from the risks associated with communal areas. As a result, there was a general decline in single women hearth-holds' livelihoods over time. An analysis of all 22 single women's pathways in Chapter 7 showed how individual livelihoods became increasingly vulnerable over their respective life courses.

8.3 Theoretical reflections

8.3.1 Access versus property, focusing on the phenomenon of *matongo*

The theoretical reflections of the thesis focus on the concepts explored in Chapter 1 in order to highlight the contributions made to existing knowledge. The study introduced the concept of access to analyse how women gain and benefit from the land resource, the concept of hearth-hold and homestead instead of household as units of analysis and the concept of livelihood decision making. The use of the concept of access versus property widened the analytical scope of the study and facilitated an appreciation of the diverse means by which people gain access to resources in the communal areas. The concept of property focuses on ownership of land sanctioned by the plural systems of rules, including the law, and not on land access for which no rules exist and where only social relations are important. In the communal areas of Zimbabwe, the legal framework governing land access does not recognise individual ownership of communal land. As a result, it would not have been possible with the property concept to explore the various extra-legal ways by which people gain access to and deal in communal area land. The concept of access provides a wider framework within which to examine how residents in communal areas gain and maintain access to land because it includes powers, and not only formal rights people have on the basis of kinship and other characteristics. One of the extra-legal land access mechanisms, *matongo*, is based on neither the market nor State-generated regulations. However, the use of *matongo* is widely accepted as a normal means of accessing land. Using a property-centered framework of analysis would have marginalised *matongo* and treated it as a temporary phenomenon yet observed practice illustrated that *matongo* is a widespread and enduring phenomenon. Whereas property focuses on mutually exclusive land claims, using the *matongo* concept has enabled the analysis of land claims as complementary rather than mutually exclusive competing activities. The terms for gaining access to *matongo* are not always clear. In some cases, the conditions are quite restrictive by, for example, denying women autonomy in deciding with whom to co-reside. However, *matongo* provided a valuable and flexible resource and relief for single women in this study who were faced with homelessness and impoverishment when they were displaced. Thus, single women's claims to *matongo* are stronger than presented in normative discussions on single women's resource access in communal areas. The *matongo* discourse shows that the male lineage has a kind of obligation to accommodate their single women. This is the main reason why women who return to their natal village are relatively more secure than women occupying *matongo* through marriage, employment or as a guardian. *Matongo* acts as a pressure valve by giving access to the desperate while retaining the claims of the absent. *Matongo* is an increasingly important way through which single women negotiate resource access in the gendered resource governance space of the communal areas. *Matongo* has enabled women to overcome the barriers imposed by the market and custom and to gain access to resources in the communal areas without directly confronting the patriarchal norms.

8.3.2 Hearth-hold versus household

The study used the concepts of hearth-hold, homestead and village to analyse the domestic unit. The selection of the hearth-hold as a unit of analysis enabled the research to focus on single women in their various marital statuses, relationships to the head of the house and relationships to men outside the formal marital relationship. Acknowledging that single women hearth-holds do not exist in a vacuum but have been shaped by their relations with other units, the homestead and village were also incorporated as units of study. Unlike the household, the hearth-hold placed the single women at the centre of the research, regardless of the individual woman's relationship with a man at any point in her life. This was useful because the male headship bias in the household unit would have confined the analysis to other male-centred ways of examining women's resource access. The woman-centred focus facilitated a continued spotlight on the dynamics of women's resource access. A household focus would have been benchmarked by the men's entry and exit from the woman's life. The use of hearth-hold rather than household challenged the notion that in patrilineal societies adult men are always the head of a domestic unit and showed that it is common for adult men to be dependents of single women. The use of the hearth-hold highlighted the autonomy and diversity of hearth-holds which the household focus would have glossed over. This is not to say that hearth-holds do not have to conform to patriarchal norms but rather that they remain in control of their individual circumstances and, in doing so, are also agents of change and decision-making units. The hearth-hold focus also showed how, contrary to popular belief, single women domestic units in a patriarchal setting can have a high degree of flexibility.

8.3.3 The concept of vulnerability

The concept of vulnerability led to an analysis of how prone individuals are to risks, how they manage the risks and what happens after the risk comes to pass. Livelihood diversification is a strategy that is commonly used to reduce vulnerability. Hearth-holds diversify by investing in non-agricultural activities and pursuing multi-spatial livelihoods. The individual women's capacity to diversify depends on their resources and decisions they make. The experiences of single women studied here show that, in a fast-changing hostile environment with multiple risks, people may become so impoverished that their capacity to respond is undermined. This highlights the impact of structural factors that produce vulnerability. For example, the existence of the communal farming areas as legal spaces creates a vulnerable population because they have no access to land outside these areas, even after the fast-track land reform (see Section 3.2.3). Communal farming areas are legal and political spaces that produce vulnerability and act as places where the ill and unemployed are dumped. The experiences of the 22 hearth-holds studied show that, even though in the short term some women successfully diversified and averted disaster, they are all destined for impoverishment in the long term as their livelihoods slide from being hearth-holds at risk to becoming succumbed hearth-holds. The at-risk hearth-holds just manage by generating enough from their daily activities to meet their immediate consumption needs. The conditions in the communal area make it difficult for hearth-holds to adequately insure themselves against subsequent shocks and long-term risks, regardless of their participating in several activities that include remittances, subsistence agriculture, wage labour, barter trade, community garden, dairy projects, beer brewing and government transfers. The diversification is not sustainable because the context continues to deteriorate. The women largely improvise and lack the resources needed to sustain the activities. The women also overextend their meagre resources – usually labour – in pursuing so many activities. This exposes them to illness. The act of diversifying is not in itself adequate to mitigate risk. There has to be sustained access to resources to enable the diversified hearth-hold to accumulate adequate stocks to insure against future risks. Without this progressive outcome, diversification becomes a futile activity that may further increase vulnerability. This research has established that, once illness afflicts a hearth-hold, there is little the domestic unit can do by way of mitigating the effects. It also emerged that no one asset was adequate to ensure livelihood security. The successful hearth-holds in the study relied on a variety of resources to ensure their survival. The use of the vulnerability concept enabled the research to show how individual hearth-holds experience the hostile environment but also how they use individual resources to confront risk. Even though individual pathways are unique, generally the hearth-holds become more impoverished, powerless and vulnerable over time, not just as a consequence of individual circumstances but also as a result of the rapid and sustained deterioration in the wider environment within which the hearth-holds reside.

8.3.4. Pathways

Through the use of pathways a concept developed to analyse decision-making in high risk environments it was possible to examine hearth-holds' decision-making and the consequences of those decisions. Even though the factors that lead to the generalised vulnerability of hearth-holds' livelihoods were identified, pathways enabled us to link these factors to individual hearth-holds' decision-making so we could understand the processes behind the decisions they took. By focusing on individual hearth-holds and having the women tell their life stories, it was also possible to capture the diversity in decision-making and outcomes that characterised the 22 single women who were the focus on the study. A focus on the individual hearth-hold experiences put those women in the centre of the stories and enabled us to capture diversity in the women's life experiences and decision-making. The study also increased our insight into how hearth-holds made decisions in the volatile conditions in Zimbabwe. The pathway reconstruction enabled us to not only illustrate the diversity of livelihood activities that the women engaged in but also to show the improvised nature of the hearth-holds' portfolios and the character of hearth-holds' livelihood altering events. As a result, it was possible to identify the main factors that resulted in hearth-holds pathways either leading them to sustainable hearth-holds or those whose livelihoods remained vulnerable to the volatile conditions and/or succumbed over their life course. The pathways enabled us to show that hearth-holds' livelihood decision-making was determined by their age, health status, marital status, specific shock to which the hearth-hold has been exposed, specific local governance conditions and health status of hearth-hold heads and rural-urban connections. It was also possible to establish the potential role of NGOs and government to influence hearth-holds' pathways. *Matongo* is also another institution that was identified as a potential factor in inducing a more progressive pathway for hearth-holds. The experiences of the 22 hearth-holds decision-making showed that kinship and networks can either increase or decrease a hearth-hold's livelihood viability. While individual cases give insights into specific experiences, it is not possible to draw generalisations at either the community or countrywide level.

8.3.5 rural-urban connections

The concept of rural-urban connections was used to analyze hearth-holds' livelihoods spatial organization. The concept facilitated a scope of analysis that extended beyond the physical boundaries of the communal area to include the urban areas, growth points and commercial farming areas where hearth-hold livelihood activities were pursued. This enabled us to expose the complimentary relationship that exists between the communal areas and urban areas in Zimbabwe and that one could not exist without the other. The rural-urban connections illustrated that the exchanges between communal areas and urban areas extended beyond the movement of people to include the exchange of goods and services. The ebbs and flows in volumes and direction of movement were influenced by how individuals responded to the various economic, social and political stimuli. The concept of connections enabled us to illustrate how the flow of goods and services between urban areas and communal areas historically evolved. The male labour market dominated rural urban mobility of the colonial era were gradually replaced by more diverse movements of both men and women as the laws and social conditions governing resource access changed over time. The evolution of rural urban connections in response to the ESAP, economic decline and government policies resulted in diminished financial flows to rural areas. People responded to this pressure by increased shorter term travel patterns, reduced investment in rural areas, splitting families between various locations and some abandoned their rural homesteads. Through the thesis, it was possible to establish how generally the evolution of rural-urban connections over time resulted in a gradual shift in the meaning of the communal area to those who had access to this land. For some the utility of the rural home as a retirement ceased after they secured freehold access to urban property while for others this was heightened. The concept of rural–urban connections enabled us to link the wider processes to local conditions and to provide steps towards answering questions about *why people in rural areas are no longer farming, why land is left fallow and why there has been a sustained deterioration of communal area homestead infrastructure over the years.*

Although rural-urban connections are diverse, they were also gendered in the study area. The gendered structuring of rural-urban connections was historically shaped by the governance framework that controlled the mobility of women and prohibited African ownership of land in urban areas. As a result, men moved to urban areas while women remained to farm and raise children. The focus on single women hearth-holds enabled us to highlight the evolution of gendered livelihood structuring across the rural-urban continuum. Single women hearth-holds secured goods and services differently than men. They had to balance the pursuit of money with their obligations to look after the young and the ill. For single women the role of the rural home as a primary residence and fall-back became increasingly important as it was here that they could negotiate access to other resources. The women's capacity to maintain homestead infrastructure, farm and even diversify their livelihood portfolio was a function of the nature of their rural-urban connections. The single women who had manages to successfully establish rural urban connections through lucrative trade, secure access to regular remittances or ownership of an urban property were best placed to maintain a viable existence in the communal areas. The women who severed or failed to establish and/or maintain a continuous flow of goods on the rural–urban continuum remained the most vulnerable as they could not produce subsistence food maintain homestead infrastructure and their livelihoods increasingly became vulnerable over time. Therefore, rural urban connections were an important determinant of livelihood sustainability.

8.4 Zimbabwe's communal farm areas, single women's land rights and the impact of AIDS: towards a more complex understanding of vulnerability

The livelihood experiences of single women are inextricably linked to the political economy of the communal farming areas in Zimbabwe. The present situation in communal areas characterised by vulnerable livelihoods and a poor resource base has historical roots. The historical reason for the establishment of communal areas as vulnerable and dependent spaces created the conditions that characterise the political economy of communal areas today. Communal farming areas were created as dependent spaces to enable the successive governments to maintain power and control over the inhabitants' access to productive resources. The land tenure regime and the location of power in traditional structures maintained land as the key productive asset under State control. As a result, even in Zimbabwe's 'good years', communal areas continued to be dependent on the State and resource flows from outside. The events of the period 1990-2003 undermined the economic linkages between the communal farming areas and the rest of the country. Since then, livelihoods in the communal areas have become more vulnerable. This has reduced the people's access to productive resources and rendered them powerless. The communal areas have changed from being a labour market able to provide their own subsistence food into a largely residential place wholly dependent on the government for sustenance. The government has used this as a strategy to foster patronage through the politicisation of food and the use of restrictive legislation to force the private sector, NGOs and other civic organisations out of the humanitarian sector. This has had the effect of concentrating power in the hands of the State and reducing the autonomy of communal areas.

With respect to the initial mission of this thesis, to establish the impact of HIV/AIDS on single women's access to land in communal areas of Zimbabwe, it can be concluded that the existing vulnerability of single women cannot be exclusively attributed to HIV/AIDS, because women's land tenure in Zimbabwe is the outcome of various factors, none of which can be conclusively studied in isolation. Women's tenure is an outcome of wide spectre of historical and socio-economic factors, including power and gender relations, rural-urban connections, misfortune, social relations and land tenure rules. HIV/AIDS is one of the many factors along with the legacy of dual development in Rhodesia, gender discrimination, governance dynamics and changing livelihoods that have shaped women's land rights. HIV/AIDS increases the vulnerability of an already vulnerable population by increasing the incidence of mortality, marital breakdown, single women hearth-hold relocation, changing the governance of resource access and eroding the labour base. A meaningful understanding of vulnerability has to embrace the other processes that impact on women's ability to gain a livelihood from the land. These include economic, social and historical circumstances. Any attempt to understand the vulnerability of women's land rights has to include these other fundamental linkages. Vulnerability therefore entails an examination not just of a livelihood's current circumstances, but also of the historical and socio-economic circumstances which the specific livelihood has experienced. Vulnerability discourse should also examine the role of the State, because commonly the State is assumed to be one of the main sources of resources to reduce risk. In Zimbabwe, the State uses vulnerability as a resource for political expediency. A sustainable solution to mediating communal areas' vulnerability, and consequently women's tenure security, should start with the restructuring of the communal areas' political economy.

References

Action Aid International (2005). Combined Harare Residents' Association, Zimbabwe Peace Project. The Impact of 'Operation Murambatsvina/Restore Order' in Zimbabwe. National Survey Report. August 2005. Available at http://www.sarpn.org.za/documents/d0001475/Impact_Operation_Murambatsvina_Aug2005.pdf, downloaded 13 January 2006.

Adams, J. (1991a). Female Wage Labour in Rural Zimbabwe. World Development 19(2/3): 163-177.

Adams, J. (1991b). The Rural Labour Market in Zimbabwe. Development and Change 22: 297-320.

Alderson, M. (1998). Women and Children First: Adjustment, Gender and Food Security in Zimbabwe. Review of African Political Economy 43: 4-23.

Alexander, J. and J. McGregor (2001). Elections, Land and the Politics of Opposition in Matabeleland. Journal of Agrarian Change 1(4): 510-533.

Aliber, M., C. Walker, M. Machera, C. Kamau, C. Omondi and C. Kanyinga. (2004). Impact of HIV/AIDS on Land Rights: Case Studies from Kenya. South Africa, HSRC.

Allen, J. (1990). Homestead Tree Planting in Two Rural Swazi Communities. Agroforestry Systems 11: 11-22.

Amin, N. (1992). State and Peasantry in Zimbabwe since Independence. European Journal of Development Research 4: 112-162.

Anderson, K. and D. Jack (1991). 'Learning to Listen' Interview Techniques and Analysis. Women's Worlds: The Feminist Practice of Oral History. D. Gluck and S. Patai (eds.). New York, Routledge: 11-26.

Andersson, J. (1999). The Politics of Land Scarcity: Land Disputes in Save Communal Area, Zimbabwe. Journal of Southern African Studies 25(4): 553-578.

Andersson, J. (2001). Re-Interpreting the Rural-Urban Connection: Migration Practices and Socio-Cultural Dispositions of Buhera Workers in Harare. Africa Development 71(1): 83-112.

Andersson, J. (2002). Sorcery in the Era of 'Henry IV': Kinship, Mobility and Mortality in Buhera District, Zimbabwe Royal Anthropological Institute 8: 425-449.

Aphane, D., N. Manzini, L. Mthembu, P. Vilikazi, T. Dhlamini and M. Magwaza (1995). Widow: Status or Description? Widowhood, Inheritance Laws, Customs and Practices in Southern Africa. N. W. and J. Stewart (eds.). Harare, WLSA: 27-57.

Armstrong, A. (1995). Rethinking Women's Law in Southern Africa. Widowhood, Inheritance Laws, Customs and Practices in Southern Africa. W. Ncube and J. Stewart (eds.). Harare, WLSA: 116-148.

Armstrong, A. (1998). Culture and Choice: Lessons from Survivors of Gender Violence in Zimbabwe. Harare, Violence against Women in Zimbabwe Project.

Ashley, C. and D. Carney (1999). Sustainable Livelihoods: Lessons from Early Experience. London, Russell Press.

Barnes, T. (1997). Am I a Man? Gender and the Pass Laws in Urban Colonial Zimbabwe, 1930-80. African Studies Review 40(1): 59-81.

Basset, M., L. Bijlmakers and D. Sanders. (1997). Professionalism, Patient Satisfaction and Quality of Health Care: Experience During Zimbabwe's Structural Adjustment Programme. Social Science Medicine 45(12): 1845-1852.

Bebbington A. (1993). Sustainable Livelihood Development in the Andes? Local Institutions and Regional Resource Use in Ecuador. Policy Review 19(1): 5-30.

References

Bebbington, A. (1999). 'Capitals and Capabilities': A Framework for Analysing Peasant Viability, Rural Livelihoods and Poverty. World Development 27(12): 2021-2044.

Bell, M. and P. Hotchkiss (1991). Garden Cultivation, Conservation and Household Strategies in Zimbabwe. Africa: Journal of the International African Institute 61(2): 202-221.

Berkvens, R. (1997). Backing Two Horses: Interaction of Agricultural and Non-Agricultural Household Activities in a Zimbabwean Communal Area. Working Paper 24. Leiden, Africa Studies Centre.

Berry, S. (1993). The Social Dynamics of Agrarian Change in Sub-Saharan Africa. Madison, University of Wisconsin Press.

Bird, K. and A. Shepherd (2003). Livelihoods and Chronic Poverty in Semi-Arid Zimbabwe. World Development 31(3): 591-610.

Blomley, N. (2004). Unsettling the City: Urban Land and the Politics of Property. London, Routledge.

Bohle, H., T. Downing and M. Watts (1994). Climate Change and Social Vulnerability. Towards a Sociology and Geography of Food Insecurity. Global Environmental Change 4(1): 37-48.

Bourque, N. (1997). Making Space: Social Change, Identity and the Creation of Boundaries in the Central Ecuadorian Andes. Bulletin of Latin American Research 16(2): 153-167.

Brand, C. (1991). Will Decentralisation Enhance Local Participation? Limits to Decentralisation in Zimbabwe: Essays on the Decentralisation of Government and Planning in the 1980s. A. Helmsing, N. Mutizwa-Mangiza, D. Gasper, C. Brand and K. Wekwete (eds.). The Hague, Institute of Social Studies: 79-98.

Bratton, M. (1987). The Comrades and the Countryside: The Politics of Agricultural Policy in Zimbabwe. World Politics 39(2): 174-202.

Bratton, M. and E. Masunungure (2006). Popular Reactions to State Repression: Operation Murambatsvina in Zimbabwe. African Affairs 1(25): 1-25.

Breslin, E. (1994). U.S. AID, the State, and Food Insecurity in Rural Zimbabwe: The Case of Gokwe. The Journal of Modern African Studies 32(1): 81-110.

Bruce, J. (1988). A Perspective on Indigenous Land Tenure Systems and Land Concentration. Land and Society in Contemporary Africa. R. Downs and P. Reyna (eds.). Hanover, University Press of New England: 23-52.

Bryceson, D. (1999). Sub-Saharan Africa Betwixt and Between: Rural Livelihoods Practices and Policies. Working Paper 43/99. Leiden, Africa Study Centre, De-Agrarianisation and Rural Employment Network.

Bryceson, D. (2000). Peasant Theories and Smallholder Policies: Past and Present. Disappearing Peasantries? Rural Labour in Africa, Asia and Latin America. D. Bryceson, C. Kay and J. Mooij (eds.). London, Intermediate Technology Publications: 1-36.

Bryceson, D. (2002). Multiplex Livelihoods in Rural Africa: Recasting the Terms and Conditions of Gainful Employment. Journal of Modern African Studies 40(1): 1-28.

Bryceson, D., C. Kay and J. Mooij, Eds. (2000). Disappearing Peasantries? Rural Labour in Africa, Asia and Latin America. London, Intermediate Technology.

Bryceson, D. and T. Mbara (2003). Petrol Pumps and Economic Slumps: Rural-Urban Linkages in Zimbabwe's Globalisation Process. Royal Dutch Geographical Society 94(3): 335-349.

Buvinic, M. and G. Gupta (1997). Female-Headed Households and Female Maintained Families: Are They Worth Targeting to Reduce Poverty in Developing Countries? Economic Development and Cultural Change 45(2): 259-280.

Cahn, M. (2002). Sustainable Livelihoods Approach. Concept and Practice. Paper presented at the 3rd Biennial Conference of the International Development Studies Network of Aotearoa New Zealand Massey University.

Cain, M. (1978). The Household Life Cycle and Economic Mobility in Rural Bangladesh. Population and Development Review 4(3): 421-438.

Carney, D. Ed. (1998). Sustainable Rural Livelihoods: What Contribution Can We Make? London, DFID.

Chadya, J. and P. Mayavo (2002). The Curse of Old Age' Elderly Workers on Zimbabwe's Large Scale Commercial Farms with Particular Reference to Foreign Farm Labourers up to 2000. Zambezia XXIX (i): 12-26.

Chambers, R. (1989). Editorial Introduction: Vulnerability, Coping and Policy. IDS Bulletin 20(2): 1-7.

Chant, S. (1997). Women-Headed Households. Diversity and Dynamics in the Developing World. New York, St Martins Press.

Chaumba, J., I. Scoones and W. Wolmer (2003). New Politics, New Politics, New Livelihoods: Agrarian Change in Zimbabwe. Review of African Political Economy 98: 585-608.

Cheater, A. (1986). The Role and Position of Women in Pre-Colonial and Colonial Zimbabwe. Zambezia XIII: 65-80.

Cheater, A. (1990). The Ideology of 'Communal' Land Tenure in Zimbabwe: Mythogenesis Enacted. Africa 60(2): 188-206.

Cheater, A. and R. Gaidzanwa (1996). Citizenship in Neo-Patrilineal States: Gender and Mobility in Southern Africa. Journal of Southern African Studies 22(2): 189-200.

Chimhowu, A. (2002). Extending the Grain Basket to the Margins: Spontaneous Land Resettlement and Changing Livelihoods in Hurungwe District, Zimbabwe. Journal of Southern African Studies 28(3): 551-573.

Chimhowu, A. and D. Hulme (2006). Livelihood Dynamics in Planned and Spontaneous Resettlement in Zimbabwe: Converging and Vulnerable. World Development 34(4): 728-750.

Chimhowu, A. and P. Woodhouse (2006a). Customary Versus Private Property Rights? Dynamics and Trajectories of Vernacular Land Markets in Sub-Saharan Africa. Journal of Agrarian Change 6(3): 346-371.

Chimhowu, A. and P. Woodhouse (2006b). Officially Forbidden but not Oppressed: Vernacular Land Markets on Communal Lands in Zimbabwe. A Case Study of Svosve Communal Lands, Zimbabwe. Paper presented at Colloque International Les Frontières de la question foncière - At the frontier of land issues. Montpellier, 2006.

Coudere, H. and S. Marijse (1991). Rich and Poor in Mutoko Communal Area. Rural Development and Planning in Zimbabwe. N. Mutizwa-Mangiza and A. Helmsing (eds.). Aldershot, Avebury: 70-91.

Cousins, B. (1990). Property and Power in Zimbabwe's Communal Lands: Implications for Agrarian Reform in the 1990s. Occasional Paper. Land Policy in Zimbabwe After Lancaster, University of Zimbabwe, Harare, Centre for Applied Social Sciences.

Cousins, B. (1993). Debating Communal Tenure in Zimbabwe. Contemporary African Studies 12(1): 29-39.

Cousins, B., D. Winer and N. Amin (1992). Social Differentiation in the Communal Lands of Zimbabwe. Review of African Political Economy 53: 5-24

De Bruijn, M. and J. van Dijk (2005). Moving People: Pathways in the Hayre-Seeno Area, Central Mali. ICCD/ASC Working Paper. Wageningen, Africa Studie Centrum.

De Haan, D. (2000). Globalisation, Localisation and Sustainable Livelihoods. European Society for Rural Sociology 40(3): 339-365.

De Haan, L. D. and A. Zoomers (2005). Exploring the Frontiers of Livelihoods Research. Development and Change 36(1): 27-47.

Dekker, M. (2004). Risk, Resettlement and Relations: Social Security in Rural Zimbabwe. Doctoral Thesis. Vrije Universiteit Amsterdam.

Dhemba, J., P. Gumbo and J Nyamusara (2002). Social Security in Zimbabwe: Phase II: *Zunde raMambo* and Burial Societies. Social Development in Africa 17(12): 132-153.

Drakakis-Smith, D. (1984). The Changing Economic Role of Women in the Urbanisation Process: A Preliminary Report from Zimbabwe. International Migration Review 18(4): 1278-1292.

Drinkwater, M. (1989). Technical Development and Peasant Impoverishment: Land Use Policy in Zimbabwe's Midlands Province. Journal of Southern African Studies 15(2): 287-305.

Drinkwater, M. (2003). HIV/AIDS and Agrarian Change in Southern Africa. United Nations Regional Interagency Co-ordination and Support Office Technical Consultation on Vulnerability in the Light of an HIV/AIDs Pandemic South Africa. Available at http://www.sarpn.org.za/documents/d0000601/Drinkwater_CARE_Sept03.pdf, downloaded 15 June 2005.

Duggan, W. (1980). The Native Land Husbandry Act of 1951 and the Rural African Middle Class of Southern Rhodesia. African Affairs 79(315): 227-239.

Eakin, H. (2005). Institutional Change, Climate, Risk and Rural Vulnerability: Cases from Mexico. World Development 33(11): 1923-1938.

Ekejiuba, F. (1995). Down to Fundamentals: Women Centred Hearth-Holds in Rural West Africa. Women Wielding the Hoe. Lessons from Rural Africa for Feminist Theory and Development Practice. D. Bryceson (ed.). London, Berg: 47-61.

Ellis, F. (1998). Household Strategies and Rural Livelihood Diversification. Development Studies 35(1): 1-38.

Ellis, F. (2003). Human Vulnerability and Food Insecurity. Policy Implications. Forum for Food Security in Southern Africa. Overseas Development Group. Available at www.odi.orguk/food-security-forum, downloaded on 6 July 2006.

Farrington, J., D. Carney, C. Ashley and C. Turton (1999). Sustainable Livelihoods in Practise: Early Application of Concepts in Rural Areas. Natural Resource Perspectives 42: 1-9.

Feeny, G. (2001). The Impact of HIV/AIDS on Adult Mortality in Zimbabwe. Population and Development Review 27(4): 771-780.

Floyd, B. (1962). Land Apportionment in Southern Rhodesia. Geographical Review 52(4): 566-582.

Folbre, N. (1988). Patriarchal Social Formations in Zimbabwe. Patriarchy and Class. African Women in the Home and the Workforce. S. Sticher and J. Parpart (eds.). Boulder, London, Westview: 61-79.

Folta, J. and E. Deck (1987). Elderly Black Widows in Rural Zimbabwe. Journal of Cross-Cultural Gerontology 2(4): 321-342

Foster, G., C. Makufa, R. Drew and E Kralovec. (1997). Factors Leading to the Establishment of Child-Headed Households: The Case of Zimbabwe. Health Transition Review 7(Supplement 2): 155-168.

Francis, E. (1992). Qualitative Research: Collecting Life Histories. Fieldwork in Developing Countries. S. Devereux and J. Hoddinott (eds.). New York, Harvester Wheatsheaf: 86-101.

Francis, E. (1998). Gender and Rural Livelihoods in Kenya. Journal of Development Studies 35(2): 72-95.

Francis, E. (2002). Rural Livelihoods, Institutions and Vulnerability in Northwest Province, South Africa. Journal of Southern Africa Studies 28(3): 531-551.

Gaidzanwa, R. (1994). Women's Land Rights in Zimbabwe. A Journal of Opinion 22(2): 12-16

Gaidzanwa, R. (1997). Non-Farm Activities and Gender in Zimbabwe. Farewell to Farms: De-Agrarianisation and Employment in Africa. D. Bryceson and V. Jamal (eds.). Aldershot, Ashgate: 157-166.

Gluckman, M. (1945). African Land Tenure. Human Problems in British Central Africa III. Lusaka, Rhodes Livingstone Institute.

Goebel, A. (2005). Gender and Land Reform. The Zimbabwe Experience. Montreal & Kingston, McGill-Queen's University Press.

Golooba-Mutebi, F. (2005). Witchcraft, Social Cohesion and Participation in a South African Village Development and Change 36(5): 937-958.

Gray, L. and M. Kevane (1999). Diminished Access, Diverted Exclusion: Women and Land Tenure in Sub-Saharan Africa. African Studies Review 42(2): 15-39.

Gregson, S., R. Anderson, J. Ndlovu, T. Zhuwau and S. Chandiwana (1997). Recent Upturn in Mortality in Rural Zimbabwe: Evidence for an Early Demographic Impact of HIV-1 Infection? AIDS 11(10): 1269-1280.

Gregson, S, T. Zhuwau, J. Ndlovu and C.A. Nyamukapa (2002). Methods to Reduce Social Desirability Bias in Sex Surveys in Low-Development Settings: Experience in Zimbabwe. Sex Transmission Disease 29: 568-575.

Griffiths, A. (1988). Support among the Bakwena. Between Kinship and the State. F. von Benda-Beckmann, K. von Benda-Beckmann, E. Casino, F. Hirtz, G. Woodman and H. Zacher (eds.). Dortrecht, The Netherlands, Foris Publications Holland: 289-315.

Griffiths, J. (1986). What is Legal Pluralism? Journal of Legal Pluralism (4): 1-55.

Gugler, J. (1989). Women Stay on the Farm No More: Changing Patterns of Rural-Urban Migration in Sub-Saharan Africa. Modern African Studies 27(2): 347-352.

Gupta, G Rao. 2000.Gender, Sexuality, and HIV/AIDS: The What, the Why and the How. Plenary Address to the XIII International AIDS Conference. Durban South Africa, July 12.

Guyer, J. and P. Peters (1987). Special Issue Introduction to Conceptualising the Household: Theory and Policy in Africa. Development and Change 18(2): 197-214.

Gysels, M., R. Pool and B. Nnalisuba (2002). Women Who Sell Sex in a Ugandan Trading Town: Life Histories, Survival Strategies and Risk. Science and Social Medicine 54: 179-192.

Haddad, L. and S. Gillespie (2001). Effective Food and Nutrition Policy Responses to HIV/AIDS: What We Know and What We Need to Know. Journal of International Development 13: 487-511.

Hansen, K., G. Woelk, H. Jackson, R. Kerkhoven, N. Manjonjori and P. Maramba (1998). The Cost of Home-Based Care for HIV/AIDS Patients in Zimbabwe. AIDS Care 10(6): 751-759.

Hartnack, A. (2005). 'My Life Got Lost': Farmworkers and Displacement in Zimbabwe. Journal of Contemporary African Studies 23(2): 173-192.

References

Helmsing, A. (1991). Non-Agricultural Enterprises in the Communal Lands of Zimbabwe. Rural Development and Planning in Zimbabwe. N. Mutizwa-Mangiza and A. Helmsing (eds.). Aldershot, Avebury: 257-276.

Hindin, M. (2002). For Better for Worse? Women's Autonomy and Marital Status in Zimbabwe. Social Science Research 31: 151-172.

Holleman, J. (1954). An Anthropological Approach to Bantu Law with Special Reference to Shona Law. Rhodes-Livingstone Journal 10: 51-64.

Holleman, J. (1969). Chief, Council and Commissioner. Some Problems of Government in Rhodesia. Leiden, Africa Studie Centrum.

Huisman, H. (2005). Contextualising Chronic Exclusion: Female-Headed Households in Semi-Arid Zimbabwe. Tijdschrift voor Economische en Sociale Geografie 96(3): 253-263.

IDS (2003). Zimbabwe Human Development Report: Redirecting Our Responses to HIV and AIDS. Harare, University of Zimbabwe.

IFPRI (1995). Divorce and Family Structure in a Changing Environment. CG Newsletter 1(3): 1-1.

Izumi, K., Ed. (2006). The Land and Property Rights of Women and Orphans in the Context of HIV and AIDS. Case Studies from Zimbabwe. Capetown, HSRC Press.

Izumi, K. (2007). Gender-Based Violence and Property Grabbing in Africa: A Denial of Women's Liberty and Security. Gender and Development 15(1): 11-23.

Izzard, W. (1985). Migrants and Mothers: Case Studies from Botswana. Journal of Southern Africa Studies 11(2): 258-280.

Jackson, C. (2003). Gender Analysis of Land: Beyond Land Rights for Women? Journal of Agrarian Change 3(4): 453-480.

Jackson, C. (2007). Resolving Risk? Marriage and Creative Conjugality. Development and Change 38(1): 109-129.

Jackson, H. and R. Kerkhoven (1995). Developing AIDS Care in Zimbabwe: A Case for Residential Community Centres? AIDS Care 7(5): 663-673.

Jackson, J. and P. Collier (1991). Incomes, Poverty and Food Security in the Communal Lands of Zimbabwe. Rural Development in Zimbabwe. N. Mutizwa-Mangiza and A. Helmsing (eds.). Aldershot, Avebury: 21-70.

Jacobs, S. (2002). Land Reform? Still a Goal Worth Pursuing for Rural Women. Journal of International Development 14: 887-898.

Jenkins, C. (1997). The Politics of Economic Policy-Making in Zimbabwe. Modern African Studies 35(4): 575-602.

Kaag, M., R. van Berkel, J. Brons, M. de Bruijn, H. van Dijk and L. de Haan (2002). Poverty is Bad: Ways Forward in Livelihood Research. Utrecht, Livelihood Group, Ceres Research School.

Kakuru, D. and G. Paradza (2007). Reflections on the Use of the Life History Method in Researching Rural African Women: Field Experiences from Uganda and Zimbabwe. Gender and Development 15(2): 287-297.

Kanji, N. (1995). Gender, Poverty and Economic Adjustment in Harare, Zimbabwe. Environment and Urbanisation 7(1): 37-56.

Kesby, M. (1996). Arenas for Control, Terrains of Gender Contestation: Guerrilla Struggle and Counter-Insurgency Warfare in Zimbabwe 1972-1980. Journal of Southern African Studies 22(4): 561-584.

Kesby, M. (1999). Locating and Dislocating Gender in Rural Zimbabwe: The Making of Space and the Texturing of Bodies. Gender Place and Culture 6(1): 27-47.

Kinsey, B. (1999). Land Reform, Growth and Equity: Emerging Evidence from Zimbabwe's Resettlement Programme. Journal of Southern African Studies 25(2): 173-196.

Kinsey, B. (2002). Survival or Growth? Temporal Dimensions of Rural Livelihoods in Risky Environments. Journal of Southern Africa Studies, 28(3): 615-629.

Kinsey, B., K. Burger and J. Gunning (1998). Coping with Drought in Zimbabwe: Survey Evidence on Responses of Rural Households to Risk. World Development 26(1): 89-110.

Kriger, N. (2005). ZANU (PF) Strategies in General Elections, 1980-2000: Discourse and Coercion. African Affairs 104(414): 1-34.

Kuper, H. (1954). The Shona. The Shona and Ndebele of Southern Rhodesia. D. Forde (ed.). London, International African Institute: 9-40.

Ladley, A. and D. Lan (1985). The Law of the Land: Party and State in Rural Zimbabwe. Journal of Southern African Studies 12(1): 88-101.

Lan, D. (1985). Guns and Rain. Guerrillas and Spirit Mediums in Zimbabwe. London, James Currey.

Lawrence, J., C. Foggin and R. Norval (1980). The Effects of War on the Control of Diseases of Livestock in Rhodesia (Zimbabwe). The Veterinary Record 107(4): 82-84.

Loewenson, R. (1991). Rural Development and Planning in Zimbabwe. Rural Development and Planning in Zimbabwe. N. Mutizwa-Mangiza and A. Helmsing (eds.). Aldershot, Avebury: 360-381.

Machingaidze, V. E. (1991). Agrarian Change from Above: The Southern Rhodesia Native Land Husbandry Act and the African Response. International Journal of African Historical Studies 3: 557-588.

Magaramombe, G. (2004). The Impact of Land Redistribution on Commercial Farm Workers. Post- Independence Land Reform in Zimbabwe. Controversies and Impact on the Economy. M. Masiyiwa (ed.). Harare, Friedrich Ebert Stiftung and Institute of Development Studies, University of Zimbabwe: 35-56.

Maluccio, J., L. Haddad and J. May. (2003). Social Capital and Gender in South Africa, 1993-1998. Household Decisions, Gender and Development. A Synthesis of Recent Research. A. Quisumbing and J. Maluccio (eds.). Washington DC, International Food Policy Research Institute: 145-152.

Marquette, C. (1997). Current Poverty, Structural Adjustment, and Drought in Zimbabwe. World Development 25(7): 1141-1149.

May, J. (1983). Zimbabwean Women in Customary and Colonial Law. Gweru, Zimbabwe, Mambo Press.

May, J. (1987). Changing People Changing Laws. Gweru, Zimbabwe, Mambo Press.

Mbiba, B. (2001). Communal Land Rights in Zimbabwe as State Sanction and Social Control: A Narrative. Africa Journal of International African Institute 71: 27-448.

McCall, M. (1996). Rural Brewing, Exclusion, and Development Policy-Making. Focus on Gender 4(3): 29-38.

McGregor, J. (2001). The Politics of Disruption: War Veterans and the Local State in Zimbabwe. African Affairs 101(402): 9-37.

Merry, S. E. (1988). Legal Pluralism. Law and Society Review 22(5): 869-896.

References

Ministry of Health and Child Welfare (2004). The HIV and AIDS Epidemic in Zimbabwe: Where Are We Now? Where Are We Going? Zimbabwe, Ministry of Health and Child Welfare, National AIDS Council, USAID.

Moore, S. (1986). Social Facts and Fabrications Customary Law on Kilimanjaro, 1880-1980. New York, Cambridge University Press.

Moore, D. (1998). Clear Waters and Muddied Histories: Environmental History and the Politics of Community in Zimbabwe's Eastern Highlands. Southern African Studies 24(2): 377-403.

Mosley, P. (1983). The Settler Economies: Studies in the Economic History of Kenya and Southern Rhodesia, 1900-1963. New York, Cambridge University Press.

Moyo, O., and S. Kawewe (2002). The Dynamics of a Racialized, Gendered, Ethnicized, and Economically Stratified Society: Understanding the Socio-Economic Status of Women in Zimbabwe. Feminist Economics 8(2): 163-181.

Mudimu, G. (2003) Zimbabwe Food Security Issues Paper. Forum for Food Security in Southern Africa. Available at www.odi.org.uk/food-security-forum, downloaded on 25 July 2007.

Mupedziswa, R. and P. Gumbo (2001). Women Informal Traders in Harare and the Struggle for Survival in an Environment of Economic Reforms. Nordic Africa Institute Report No 117. Uppsala, Sweden, Nordic Africa Institute (NAI).

Murray, C. (1980). Migrant Labour and Changing Family Structure in the Rural Periphery of Southern Africa. Journal of Southern African Studies 6(2): 139-156.

Murray, C. (1987). Class, Gender and the Household: The Developmental Cycle in Southern Africa. Development and Change 18: 235-249.

Murray, C. (2001). Livelihoods Research: Some Conceptual and Methodological Issues. Chronic Poverty Research Centre Working Paper. Available at www.chronic poverty.org/p/477/publication-details.php, downloaded on 22 May 2004.

Mutangadura, G. (2000). Household Welfare Impacts of Mortality of Adult Females in Zimbabwe: Implications for Policy and Program Development. Paper presented at AIDS and Economics Symposium organised by IAEN Network, Durban, South Africa.

Muzvidziwa, V. (2000). Food Vending: Adaptation Under Difficult Circumstances. Social Development in Africa 15(2): 69-91.

Muzvidziwa, V. (2001). Marriage as a Survival Strategy: The Case of Masvingo, Zimbabwe. Zambezia XXXVIII (II): 147-165.

Muzvidziwa, V. (2002). An Alternative to Patriarchal Marriage: Mapoto Unions. Nordic Journal of African Studies 11(1): 138-155.

Mvududu, S. and P. McFadden (2001). Reconceptualising the Family in a Changing Southern African Environment. Harare, Women and Law in Southern Africa Research and Education Trust.

NACP (1998). HIV/AIDS in Zimbabwe: Background Projections Impact and Interventions. Harare, National Aids Control Programme/Ministry of Health, Zimbabwe Government.

Ncube, (2004). Support for the Land Reform Programme, Availability of Inputs and Pricing System. Paper presented at the pre-budget seminar October 2004, Mutare, Zimbabwe.

Norton, A. (1983). Family Life Cycle: 1980 Journal of Marriage and Family 45(2): 267-275.

Nyambara, P. (2001a). The Politics of Land Acquisition and Struggles Over Land in the 'Communal' Areas of Zimbabwe: The Gokwe Region in the 1980s and 1990s. Africa 71(2): 253-285.

Nyambara, P. (2001b). Closing the Frontier: Agrarian Change, Immigrants and the Squatter Menace in Gokwe Region in the 1980s and 1990s. Journal of Agrarian Change 1(4): 534-549.

Odgaard, R. and A.W. Bentzon (2007). Rural Women's Access to Landed Property. Unearthing the Realities Within an East African Setting. Human Rights, Plural Legalities and Gendered Realities. Paths Are Made by Walking. A. Hellum, J. Stewart, S. Ali and A. Tsanga (eds.). Harare, Southern and Eastern African Regional Centre for Women's Law University of Zimbabwe with Weaver Press: 202-235.

O'Flaherty, M (1998). Communal Tenure in Zimbabwe: Divergent Models of Collective Land Holdings in the Communal Areas. Africa 68(4): 537-577.

O'Laughlin, B. (1995). Myth of the African Family in the World of Development. Women Wielding the Hoe. Lessons from Rural Africa for Feminist Theory and Development Practise. D. Bryceson (ed.). Oxford, Berg: 63-92.

O'Laughlin, B. (1998). Missing Men? The Debate over Rural Poverty and Women-Headed Households in Southern Africa Journal of Peasant Studies 25(2): 1-48.

Page, S. and H. Page (1991). Western Hegemony over African Agriculture in Southern Rhodesia and its Continuing Threat to Food Security in Independent Zimbabwe. Agriculture and Human Values Fall: 1-18.

Palmer, R. (1990). Land Reform in Zimbabwe, 1980-1990. African Affairs 89: 163-181.

Pankhurst, D. (1991). Constraints and Incentives in 'Successful' Zimbabwean Peasant Agriculture: The Interaction between Gender and Class. Journal of Southern African Studies 17(4): 611-632.

Paradza, G. (2007). Losers, Weepers: An Analysis of the Low-Income Widows' Experiences in Inheriting Immovable Urban Property in Zimbabwe. Poster Presentation at the Urban Research Symposium 2007. Washington DC, World Bank.

Paradza, G. (2008) *Le Esperience delle vedove nell'edita delle abitazioni urbane in Zimbabwe* Afriche e oriente 3-4:64-80

Paradza G, (2009) Intergenerational Struggles over Urban Housing: the impact on livelihoods of the Elderly in Zimbabwe. Gender and Development 17, (3): 417-425.

Pederson, P. (1997). Small African Towns, Between Rural Networks and Urban Hierarchies. Aldershot, Avebury.

Peters, P. (1995). Manoeuvres and Debates in the Interpretation of Land Rights in Botswana. Africa Journal of the International African Institute 62(3): 413-434.

Peters, P. (2004). Inequality and Social Conflict over Land in Africa. Journal of Agrarian Change 4(3): 269-314.

Phimister, I. (1993). Rethinking the Reserves: Southern Rhodesia's Land Husbandry Act Reviewed. Journal of Southern African Studies 19(2): 225-239.

Pollak, B. (1975). Black Farmers and White Politics in Rhodesia. African Affairs 74(296): 263-277.

Potts, D. (1995). Shall we go home?' Increasing Urban Poverty in African Cities and Migration Processes. The Geographical Journal 161(3): 245-264.

Potts, D. (2000a). Worker-Peasants and Farmer-Housewives in Africa: The Debate about 'Committed' Farmers, Access to Land and Agricultural Production. Journal of Southern African Studies 26(4): 807-832.

Potts, D. (2000b). Urban Unemployment and Migrants in Africa: Evidence from Harare 1985-1994. Development and Change 31: 879-910.

Potts, D. (2006). All My Hopes and Dreams Are Shattered': Urbanisation and Migrancy in an Imploding African economy- The Case of Zimbabwe. Geoforum 37(4): 536-551.

Potts, D. and C. Mutambirwa (1990). Rural-Urban Linkages in Contemporary Harare: Why Migrants Need their Land. Journal of Southern African Studies 16(4): 677-698.

Potts, D. and C. Mutambirwa (1998). 'Basics Are Now a Luxury'; Perceptions of Structural Adjustment's Impact on Rural and Urban Areas in Zimbabwe. Environment and Urbanisation 10(1): 55-76.

Quisumbing, A. (2003). What Have We Learned from Research on Intra-Household Allocation? Household Decisions, Gender and Development. Washington DC, International Food Policy Research Institute: 1-16.

Rakodi, C. (1995). The Household Strategies of the Urban Poor: Coping with Poverty and Recession in Gweru, Zimbabwe Habitat International 19(4): 447-471.

Ranga, R. (2004). The Migration Effects of Economic Structural Adjustment Programmes on a Rural Community in Zimbabwe. African Population Studies 19(1): 165-185.

Ranger, T. (1982). Tradition and Travesty: Chiefs and the Administration in Makoni District, Zimbabwe, 1960-1980. Africa: Journal of the International African Institute 52(3): 20-24.

Ranger, T. (1985). Peasant Consciousness and Guerrilla War in Zimbabwe: A Comparative Study. London, James Currey.

Ranger, T. (1993). The Communal Areas of Zimbabwe. Land in African Agrarian Systems. T. Basset and D. Crummey (eds.). Wisconsin, University of Wisconsin Press: 354-385.

Razavi, S. (2003). Introduction: Agrarian Change, Gender and Land Rights. Agrarian Change 3(1-2): 2-32.

Ribot, C. and N. Peluso (2003). A Theory of Access. Rural Sociology 68(2): 153-181.

Rocheleau, D. (1995). Maps, Numbers, Text and Context: Mixing Methods in Feminist Political Ecology. Professional Geographer 47(4): 458-466.

Rose, C. (1994). Property and Persuasion. Essays on the History, Theory and Rhetoric of Ownership. Boulder, Westview Press.

Rukuni, M. and C. Eicher (1987). The Food Security Equation in Southern Africa. Michigan, Michigan State University.

Runyan, W. (1982). The Life History: Hub of History, Psychology, and Sociology. Life Histories and Psychobiography: Explorations in Theory and Method. New York, Oxford University Press: 187-191

Russell, M. (1993). Are Households Universal? On Misunderstanding Domestic Groups in Swaziland. Development and Change 24: 755-785.

Rutherford, B. (1997). Another Side to Rural Zimbabwe: Social Constructs and the Administration of Farm Workers in Urungwe District, 1940s. Journal of Southern African Studies 23(1): 107-126.

Rutherford, B. (2001). Farm Workers in Zimbabwe. Working on the Margins: Black Workers, White Farmers in Post Colonial Zimbabwe. London, Zed.

Rwezaura, B. (1989). Changing Community Obligations to the Elderly in Contemporary Africa. Social Development in Africa 4(1): 5-24.

Rwezaura, B., A. Armstrong, W. Ncube, J. Stewart, P. Letuka and P. Mosonya (1995). Parting the Long Grass: Revealing and Reconceptualising the African Family. Journal of Legal Pluralism and Unofficial Law 35: 25-73.

Sachikonye, L. (2003). The Situation of Commercial Farmworkers after Land Reform in Zimbabwe. A Report for the Farm Community Trust of Zimbabwe. London, Catholic Institute for International Relations.

Schmidt, E. (1992). Peasants, Traders, and Wives. Shona Women in the History of Zimbabwe, 1870-1939. Harare, Baobab.

Schott, R. (1988). Traditional Systems of Social Security and their Present-Day Crisis in West Africa. Between Kinship and the State. Social Security and Law in Developing Countries. F. von Benda-Beckmann, K. von Benda-Beckmann, E. Casino, F. Hirtz, G. Woodman and H. Zacher (eds.). Dortrecht, The Netherlands, Foris Publications Holland: 89-108.

Scoones, I. (ed.). (1995) Living with Uncertainty: New Directions for Pastoral Development in Africa. London, Intermediate Technology Publications.

Scoones, I., C. Chinaniso, S. Chikura, S, Jeranyama, P. Machaka and P Mavedzenge (1996). Hazards and Opportunities. Farming Livelihoods in Dryland Africa: Lessons from Zimbabwe. London, Zed.

Scoones, I. and W. Wolmer (2003). Part 1: Contexts and Debates. Introduction: Livelihoods in Crisis: Challenges for Rural Development in Southern Africa. IDS Bulletin 34(3): 1-14.

Sen, A. (1981). Poverty and Famines. Oxford, Oxford University Press.

Sharp, J. and A. Spiegel (1985). Vulnerability to Impoverishment in South African Rural Areas: The Erosion of Kinship and Neighbourhood as Social Resources. Africa 55(2): 133-152.

Shipton, P. and M. Goheen (1992). Understanding African Land-Holding: Power, Wealth and Meaning. Africa 62(3): 307-325.

Shutt, A. (2002). Squatters, Land Sales and Intensification in Marirangwe Purchase Area, Colonial Zimbabwe, 1931-65. Journal of African History 43: 473-498

Sithole, B., B. Campbell, D. Dore and W. Kozanayi (2003). Narratives on Land: State-Peasant Relations over Fast Track Land Reform in Zimbabwe. African Studies Quarterly 7(2&3). Available at http://web.africa.ufl.edu/asq/v7/v7i2a4.htm, downloaded on 15 June 2005

Solway, J. (1994). Drought as a 'Revelatory Crisis': An Exploration of Shifting Entitlements and Hierarchies in the Kalahari, Botswana. Development and Change 25(3): 471-495.

Thompson, G. (2004). Cultivating Conflict: Agricultural 'Betterment', The Native Land Husbandry Act (NLHA) and Ungovernability in Colonial Zimbabwe, 1951-1962. Africa Development XXIX (3): 1-39.

Thorner, D., B. Kerblay and R. Smith (1966). AV Chayanov on the Theory of Peasant Economy. Illinois, Irwin Inc.

Tremblay (1957). Key Informant Technique: A Non-Ethnographic Application. American Anthropologist 59(4): 699-701.

Udry, J., N. Dole and K. Gleiter (1992). Forming Reproductive Unions in Urban Zimbabwe. International Family Planning Perspectives 18(1): 10-12.

UN (2005). Report of the Fact-Finding Mission to Assess the Scope and Impact of Operation Murambatsvina by the UN Special Envoy on Human Settlements Issues in Zimbabwe. Mrs Anna Tibaijuka. Available at http://www.un.org/News/dh/infocus/zimbabwe/zimbabwe_rpt.pdf, downloaded on 13 January 2006.

UNAIDS (1998). Zimbabwe National HIV and AIDS Estimates, 1998. Harare, Zimbabwe Ministry of Health and Child Welfare.

UNAIDS (2003). Zimbabwe National HIV and AIDS Estimates, 2003. Harare, Zimbabwe Ministry of Health and Child Welfare.

UNAIDS (2005). Zimbabwe National HIV and AIDS Estimates, 2005. Harare, Zimbabwe Ministry of Health and Child Welfare.

UNDP (2004a). Zimbabwe Millennium Development Goals: 2004 Progress Report. Harare.

UNDP (2004b). Zimbabwe Human Development Report 2003: Redirecting Our Responses to HIV/AIDS. Harare, Poverty Reduction Forum and Institute of Development Studies.

Varley, A. (1996). Women Heading Households: Some More Equal than Others? World Development 24(3): 505-520.

Van Dijk, J, 2002. Book Review of Mortimore and Adams. Working the Sahel. Environment and Society in Northern Nigeria. Development and Change 33(3): 565-566.

Vijfhuizen, C. (2002). The People You Live With: Gender Identities and Social Practices, Beliefs and Power in the Livelihoods of Ndau Women and Men in a Village with an Irrigation Scheme in Zimbabwe. Harare, Weaver Press.

Walker, C. (2002). Land Reform in Southern and Eastern Africa: Key Issues for Strengthening Women's Access to and Rights in Land, Study Commissioned by FAO. Sub-Regional Office for Eastern and Southern Africa. Harare, Zimbabwe, FAO.

Walker, L. (2006). Women Without Property: Female Farm Workers in Zimbabwe. Reclaiming Our Lives: HIV and AIDS, Women's Land and Property Rights and Livelihoods in Southern and East Africa: Narratives and Responses. K. Izumi (ed.). Capetown, HSRC Press: 109-111. Available at www.hsrcpress.ac.za, downloaded on 15 September 2007.

Watts, M. and H. G. Bohle (1993). Hunger, Famine and the Space of Vulnerability GeoJournal 30(2): 117-125.

Weinrich, A. (1975). African Farmers in Rhodesia. Old and New Peasant Communities in Karangaland. London, Oxford University Press.

Wekwete, K. (1991). Growth Centre Policy in Zimbabwe with Special Reference to District Service Centres. Rural Development in Zimbabwe. N. Mutizwa-Mangiza and A. Helmsing (eds.). Aldershot, Avebury: 187-221.

Whitehead, A. (2004). Persistent Poverty in Upper East Ghana. Basis Brief. Collaborative Research Support Program No 26, November 2004: 1-4.

Whitehead, A. and D. Tsikata (2003). Policy Discourses on Women's Land Rights in Sub-Saharan Africa: The Implications of the Return to Custom. Journal of Agrarian Change 3(1-2): 67-112.

Wisner, B., P. Blaikie, T. Cannon and I. Davis (2004). At Risk: Natural Hazards, People's Vulnerability and Disasters. London, Routledge.

WLSA (1990). The Legal Situation of Women In Southern Africa. J. Stewart and A. Armstrong (eds.). Mount Pleasant, Harare, University of Zimbabwe Publications.

WLSA (1992). Struggling Over Scare Resources: Women and Maintenance in Southern Africa. (Regional Report Phase One). Harare, WLSA Regional Office. (1992, reprinted 1997).

WLSA (1994). Inheritance in Zimbabwe: Law, Customs and Practices. Zimbabwe, WLSA (1994, revised 1996).

WLSA (1995). Widowhood, Inheritance Laws, Customs and Practices in Southern Africa: Regional Comparative Report on Inheritance. W. Ncube and J. Stewart (eds.). Harare, WLSA.

WLSA (1996). Regional WLSA Impact Study Report. Harare, WLSA.

WLSA (1997). Paving a Way Forward: A Review and Research Primer of WLSA Research Methodologies. J. Stewart (ed.). Harare, Bardwell Printers.

WLSA (2002) A Reflection: Women's Access to land in existing land tenure systems in the WLSA countries. WLSA Research and Educational Trust, Harare

Wolmer, W. and Scoones, I. (2000). The Science of 'Civilised' Agriculture: The Mixed Farming Discourse in Zimbabwe. African Affairs 99: 575-600.

Yamano, T. and T. Jayne (2004). Measuring the Impacts of Working Age Adult Mortality on Small-Scale Farm Households in Kenya. World Development 23(1): 91-119.

Yamin, F., R. Atiq and H Saleemul (2005). Vulnerability, Adaptation and Climate Disasters: A Conceptual Overview. IDS Bulletin 36(4): 1-14

Yaro, J. (2006). Is Deagrarianisation Real? A Study of Livelihood Activities in Rural Northern Ghana. Modern African Studies 44(1): 125-156.

Yin, R. (2003). Case Study Research: Design and Methods. Thousand Oaks, Sage.

Zinyama, L. (1991). Agricultural Development Policies in the African Farming Areas of Zimbabwe. Rural Development and Planning in Zimbabwe. N.D. Mutizwa-Mangiza and A.H. Helmsing (eds.). Aldershot, Avebury: 95-112.

Summary

Single women, land and livelihood vulnerability in a communal area in Zimbabwe

This thesis focused on the ways in which single women take decisions to cope with risk and gain access to resources in communal areas of Zimbabwe. A study into the experiences of single women hearth-holds was necessary because there was a dearth of systematic research on how these women accessed resources in a patriarchal and legal space. Communal areas have been undergoing significant changes as a result of government policies, HIV/AIDS, erratic weather, ambiguous governance, the collapse of the agricultural economy, hyperinflation and the general deterioration of the Zimbabwean economy over the last decade. This has resulted in changing rural-urban connections, increasing poverty, depleted resource base, a high death rate, loss of traditional sources of cash income, a decline in the capacity to produce or procure subsistence food and an uncertain decision-making environment. This study has established that, although resource governance systems in communal farming area are male biased, within those systems, there are more diverse opportunities of access than previously thought and single women have more room to manoeuvre than previously thought. The norms used to determine single women's access to resources are flexible and negotiable. However, the flexibility of the communal area resource governance has increased both the opportunities and the vulnerabilities of single women's residence access in the communal areas. This paradox has arisen because of the peculiar dynamic of different processes that influence resource access in these areas and the specific circumstances of single women at any one time.

This research was initially designed to document the ways in which gender inequality in relation to control of land made single women particularly vulnerable to impoverishment and the impact of HIV/AIDS in Zimbabwe. The research context - Zimbabwean communal areas - has been undergoing rapid changes in the past two decades, which increased the vulnerability of communal area livelihoods by undermining the communal area economy. The people's reactions to the increasing risks induced changes in relationships governing resource access in the communal areas. There is a large body of literature on gendered land rights of rural women in Southern Africa which suggests that the secondary land rights of these women leave single women vulnerable to poverty and destitution, particularly in a time of HIV/AIDS. This thesis suggests that the reality is more complex in rural Zimbabwe: firstly because gendered land rights are not well captured by the distinction between primary and secondary rights, and secondly because rural livelihoods are no longer just dependent on land.

Instead of focusing only on customary land tenure and property relations, this research focused on a broader concept of land access to analyse the tenure position of women. The research looked at decision making by single women in the face of the multiple challenges emanating from unpredictable economic, political and social conditions and rapid changes in domestic units. The livelihood decision making of single women is analysed through the concepts of access and hearth-holds, which put women at the centre of the research. The focus on hearth-holds as a unit of analysis represented a major departure from the conventional domestic units used in patrilineal land research. The concept of access focuses our attention not only on property rights in land, but also at the ability of people to benefit from resources through various mechanisms and powers they have. The livelihood framework was used for looking at different kinds of resources along with the related concept of vulnerability. The pathways concept developed to study decision making in risky areas, which focuses on the improvised character of many livelihood decisions was used to analyse livelihood decision making and to illuminate the diversity of livelihood activities and portfolios of single women.

The aim of the research was to find out how in this period of heightened vulnerability, single women took livelihood decisions and gained access to productive resources in communal farming areas of Zimbabwe? The question was divided into four sub questions:

1. How have historical changes in the wider environment of Zimbabwe affected resource access and livelihood decision making in the communal areas of Zimbabwe?
2. What are the emergent modes of access to land and other resources through which women and men construct their livelihoods in the rapidly changing communal areas of Zimbabwe?
3. What are the specific vulnerabilities confronted by single women in obtaining access to land for residential and agricultural purposes in the communal areas in the context of these changes in the wider environment?
4. How do single women organise livelihoods and take livelihood decisions in the communal areas of Zimbabwe?

The research fieldwork was conducted between August 2004 and December 2005 in two Shona villages in the communal areas of Zimbabwe. Chapter 2 describes the research design strategy, research experience and the data collection methods employed. The chapter shows how the research design was adapted to accommodate observed differences between the literature studied and ethnographic reality in the field, the author's considerations in selecting the research area, villages and units of analysis. The chapter, through a description of the researcher's experiences, demonstrates the shortcomings of normative frameworks for investigating single women hearth-holds' decision making. It also illustrates the theoretical, methodological and practical challenges of using conventional research methodologies in such a fast-changing environment and how new methods were devised. The field experiences show that fast-changing environments require a sensitive and flexible approach. The research site was selected because of security concerns, a crippling fuel crisis in Zimbabwe at the time and the opportunity the site offered to study livelihood diversification. The research focused on 22 single women hearth-holds, homestead, village and ward. The use of multiple units of analysis was necessary because of the importance of linkages between the hearth-holds and other socio-economic spaces such as urban areas.

Chapter 3, based on secondary data, explores the historical and contextual framework of land tenure and livelihood decision making of single women in communal farming areas. The chapter constructs a framework within which to analyse how single women take livelihood decisions and gain access to productive resources in the communal areas. The complexity of single women's land rights and the vulnerability of their livelihoods need to be understood historically, taking account both of long-term transformations in the colonial period and of the immediate history of political turbulence, economic stagnation and AIDS in Zimbabwe. The transformation of institutions in communal areas governing resource access has always been contingent upon larger economic and political developments in the country. This Chapter shows how this created all kinds of constraints and opportunities for livelihoods in communal areas for the population at large as well as single women. The chapter also highlights how the legal complexity introduced by successive governments created a diversity of institutional spaces which have shaped the resource access of single women and other people in the communal areas.

Chapter 4 describes the specific spatial and historical context of the study. The chapter shows how the rapid changes experienced by the communal areas in the last two decades increased livelihood differentiation and altered gendered relations. The impact of the rapid changes in each communal area was mediated by the location, resource endowment and local governance context. The impact of the shocks on the different households and people was mediated by their respective positions. The study of two different villages, Ndamba and Makuku, located in the same ward demonstrates how different modes of local governance influenced resource access and increased the diversity of outcomes for single women in the same communal area. The chapter, drawing from the village inventory also highlights how the communities within which the single women were located experienced the deterioration of communal areas and processes described in Chapter 3.

Chapter 5 looks at how single women negotiated community membership and access to homesteads. The chapter highlights the importance for vulnerable women of access not just to land as a resource but also to membership of the community in which the land is located. The chapter shows that single women gain access to residence in communal areas in a very uncertain environment. The findings of this study are that single women have more room for manoeuvre in deciding where to set up their residence. The chapter shows how changes in marital status wrought by death or divorce alter the conditions upon which the women maintain residential and community access. The resultant changes in terms of access mean that women have to resume negotiations with different people who assume control of the homestead and community membership. The multiplicity of variables that influence single women's residence access is inconsistent with findings that infer that customary tenure systems generally result in the eviction of widows and restrict hearth-holds' access to residence in their natal homestead. Hearth-holds deserted homesteads, were nudged out by the community or were pulled out by their natal relatives. These tendencies are neither the dominant nor the only outcomes of single women's residence status following the change of marital status. There was no single outcome of these negotiations as each single woman's situation was unique. It appears vulnerability of single women hearth-holds' residence in communal areas is linked to the vulnerability of their children's residential rights. The importance of support and social relations for single women cannot be overstated. There is no general impact of the various factors on single women's residence. The conditions may interlock to produce extreme vulnerability, but the reverse is also true. However, access to residential land remains subject to constant negotiation. By exploring the strategies and paradoxes, the chapter shows that women are not only passive recipients but also active decision makers who influence the outcomes following changes in their marital status.

Chapter 6, explains the tragic puzzle of agricultural production in rural communal areas today – the presence of uncultivated land in the midst of scarcity and extreme poverty – and shows how hearth-holds' access to agricultural land fits within it. A focus on practices and individual single women hearth-holds rather than norms highlights an increasingly important mechanism through which people gain and maintain access to land and residence in the communal areas of Zimbabwe. The chapter shows how indeterminate governance institutions, migration, and high morbidity and mortality processes intersect to produce diverse resource access opportunities and vulnerabilities for single women and other marginalised people in the communal areas. Various extra-legal methods that single women use to access resources and to reduce their vulnerability in communal areas are described. These opportunities are known as *matongo (vacant* lands*)*. Through the concept of *matongo* the chapter deconstructs the meaning of 'vacant land' in the context of the communal areas. It illustrates how the processes of migration, mortality and legal ambiguity identified in Chapter 3 nurture *matongo*. The chapter shows how single women gain resource access through the *matongo* layer and identifies the vulnerabilities of these opportunities. The terms for gaining access to *matongo* are not always clear. In some cases, the conditions are quite restrictive by, for example, denying women autonomy to decide with whom to co-reside. *Matongo* act as pressure valve by giving access to the desperate, while retaining the claims of the absent. The indeterminate rules governing *matongo* are a source of uncertainty for people who gain access to land through *matongo.* This is a source of vulnerability. The unclear governance of *matongo* results in some people retaining ownership, while others dispose of their land claims on out-migration. *Matongo* enable women to overcome the barriers imposed by the market and custom, and to gain access to resources in the communal areas without directly confronting the patriarchal norms. This examination of *matongo* has enabled the analysis of land claims as complementary rather than mutually exclusive and competing categories.

In Chapter 7, livelihood decision making of the 22 hearth holds as examined through their life stories shows how single women made livelihood decisions in the fast-changing environment. The pathways of the single women in this study illustrate the limited opportunities for diversification available to the increasingly vulnerable women. Each of the 22 cases was unique. The extent to which single women could diversify their livelihoods depended on the hearth-hold's assets, health of members, developmental stage of hearth-holds, access to non-agricultural income, access to residence on a homestead, local governance and the specific shocks to which the hearth-hold was exposed. The continued exposure of livelihoods to shocks further depleted the resource base and limited the diversification options. The portfolios of single women reflected the reduced importance of subsistence agriculture because of the increasing risk associated with agriculture and single women hearth-holds' limited capacity to mobilise the labour and agricultural inputs needed to engage in agriculture. The chapter shows the difficult circumstances of these single women which resulted in some of them undertaking risky activities. The chapter also shows how the local context within the two study villages influenced opportunities for diversification and how hearth-holds spread their activities over different locations, extending to growth points, former commercial farming areas, urban areas and neighbouring countries to cope with the limits imposed by the communal area environment and to reduce risk. The single women's capacity to explore multi-spatial livelihoods was limited by the presence of dependent children, the women's age, the health of the hearth-hold's members and the women's limited labour skills. Generally, the combination of localised and distant diversification options failed to protect women from the risks associated with communal areas. As a result, there was a general decline in the viability of single women hearth-holds' livelihoods over time.

Chapter 8 concludes the thesis by drawing together all the discussions reflect on lessons learnt and map out the contribution of this thesis to existing knowledge. This study has established that, although communal area resource governance systems are male biased, within those systems, there are more diverse forms of access than previously thought. The different ways in which single women access land, from allocation of primary rights to the possibility of divorced women maintaining their marital residence, reveal that anything is possible because existing rules are manipulated, contradictory and used strategically by, for example, the village heads. The thesis showed how single women's rights remain hidden in investigations that emphasise the use of formal resource access frameworks to analyse land access. The formal frameworks dismiss or deny the importance of these so-called secondary rights. Women's land claims are not handled in the same way as men's, but can be enforced socially and customarily. These claims, though weaker than men's, are significant as single women cannot be thrown off the land, and the women can use these claims as a negotiating base for access to other resources, such as assistance with children, illness and livelihood organisation. The finding that widows who remained in their marital residence after the death of their spouse constitute the majority of the single women in this study highlights the importance of the marital relationship for women's access to residence. The finding that divorced women were generally accommodated among their birth kin emphasises the importance of the natal relationship for divorced single women's land access. Even though the kinship-based modes of land access remain important for single women, new and innovative ways have been introduced through the market, employment and derived rights by dependents of single women.

Studies of women's land-right vulnerability in the developing world overemphasise the role of patriarchy. The thesis illustrates that patriarchy is only one of several institutions governing resource access in Zimbabwe's communal areas. This is because of the existence of a variety of governance structures and institutions which draw power from a variety of sources, including the government, the dominant political party, traditional authorities and formal legislation. The diversity of authorities, lack of clear and systematic rules, the decentralisation of the decision-making process from the wider community to individual families and the diversity introduced by in-migration and marketing of communal area homesteads have rendered legal systems and rules irrelevant in determining resource access in communal areas. This has provided opportunities for people to negotiate resources through other institutions and arrangements.

Through the use of two villages located in the same ward the thesis demonstrated the mediating effect of the local governance context and increased the diversity of outcomes for single women in the same communal area. The respective *sabhuku*'s position on the land market and the resulting extreme outcomes for single women and the general population in the two villages highlights not only the role of local authorities but also the importance of the specific local context in shaping resource access. The thesis showed that the differential interpretation of the regulations governing the land market in communal areas combined with the other processes to produce significant differences in the dynamics of single women's resource access. The existence of a land market in one village increased population, opportunities for accessing land and joining the community and increased opportunities for vulnerable single women to secure their livelihoods through exploiting opportunities on the land market. In the same village, the market reduced single women's land access through kinship, increased competition for scarce transfers from the State, undermined kinship support and increased the vulnerability of single women who controlled underutilised land as the men in power took advantage of the land market. The refusal by the *sabhuku* in the other village to facilitate a land market secured single women's access to communal land as a fall back, reduced the strain on the community by restricting the immigration rate, increased the community and consequently single women's capacity to benefit from State and NGO transfers and enhanced community cohesion. On the other hand, this restriction of the land market undermined single women's opportunity to diversity and benefit from the land market, denied single women autonomy in terms of decision-making about their children and left the women's land access vulnerable to opportunistic behaviour by the *sabhuku*. The thesis showed that while the two localities exhibited differences that impacted on single women's livelihood decision-making, the local context had no impact on mediating women's vulnerability to HIV/AIDS.

The livelihood decision making of single women examined through their life stories shows how single women managed in the fast-changing environment. The study found that, although the importance of land as an agricultural asset had declined for single women in the communal areas, access to residential land and the community is still the basis for negotiating access to social security and livelihood opportunities. However, this in itself did not guarantee secure livelihoods.

The study contributes to knowledge by introducing the concept of access to analyse how women gain and benefit from the land resource, the concept of hearth-hold and homestead instead of household as units of analysis and the concept of livelihood decision making. The use of the concept of access versus property widened the analytical scope of the study and facilitated a better appreciation of the diverse means by which people gain access to resources in the communal areas. The concept of property focuses on ownership of land sanctioned by the plural systems of rules, including the law, and not on land access for which no rules exist and where only social relations are important. In the communal areas of Zimbabwe, the legal framework governing land access does not recognise individual ownership of communal land. As a result, it would not have been possible with the property concept to explore the various extra-legal ways by which people gain access to and deal in communal area land. The concept of access provides a wider framework within which to examine how residents in communal areas gain and maintain access to land because it includes powers, and not only formal rights people have on the basis of kinship and other characteristics. One of the extra-legal land access mechanisms, *matongo*, is based on neither the market nor State-generated regulations. However, the use of *matongo* is widely accepted as a normal means of accessing land. Using a property-centered framework of analysis would have marginalised *matongo* and treated it as a temporary phenomenon yet observed practice illustrated that *matongo* is a widespread and enduring phenomenon. Whereas property focuses on mutually exclusive land claims, using the *matongo* concept has enabled the analysis of land claims as complementary rather than mutually exclusive categories. *Matongo* has enabled women to overcome the barriers imposed by the market and custom and to gain access to resources in the communal areas without directly confronting the patriarchal norms.

The study used the concepts of hearth-hold, homestead and village to analyse the domestic unit. The selection of the hearth-hold as a unit of analysis enabled the research to focus on single women in their various marital statuses, relationships to the head of the house and relationships to men outside the formal marital relationship. The hearth-hold placed the single women at the centre of the research, regardless of the individual woman's relationship with a man at any point in her life. This was useful because the male headship bias in the household unit would have confined the analysis to other male-centred ways of examining women's resource access. The woman-centred focus facilitated a continued spotlight on the dynamics of women's resource access. A household focus would have been benchmarked by the men's entry and exit from the woman's life. The use of hearth-hold rather than household challenged the notion that in patrilineal societies adult men are always the head of a domestic unit and showed that it is common for adult men to be dependents of single women. The use of the hearth-hold highlighted the autonomy and diversity of hearth-holds which the household focus would have glossed over. This is not to say that hearth-holds do not have to conform to patriarchal norms, but rather that they remain to a certain extent in control of their individual circumstances and, in doing so, are agents of change. The hearth-hold focus also showed how, contrary to popular belief, single women domestic units in a patriarchal setting can have a high degree of flexibility.

The concept of vulnerability focuses our attention on how prone individuals are to risks, how they manage the risks and what happens after the risk comes to pass. Livelihood diversification is a strategy that is commonly used to reduce vulnerability. The experiences of single women studied here show that, in a fast-changing hostile environment with multiple risks, people may become so impoverished that their capacity to respond is undermined. The experiences of the 22 hearth-holds studied show that, even though in the short term some women successfully diversified and averted disaster, they are all destined for impoverishment in the long run as their livelihoods slide from being hearth-holds at risk to becoming succumbed hearth-holds. This showed that the mere act of diversifying is not in itself adequate to mitigate risk. This research has established that, once illness afflicts a hearth-hold, there is little the domestic unit can do by way of mitigating the effects. It also emerged that no one asset was adequate to ensure livelihood security.

The research was originally oriented at establishing the impact of HIV/AIDS on single women's access to land in communal areas of Zimbabwe, concluded that the existing vulnerability of single women cannot be exclusively attributed to HIV/AIDS, because women's land tenure in Zimbabwe is the outcome of various factors, none of which can conclusively be studied in isolation. Women's tenure is an outcome of a multitude of historical and socio-economic developments, including power and gender relations, misfortune, social relations and land tenure rules. HIV/AIDS is one of the many factors along with the legacy of dual development in Rhodesia, economic crisis, gender discrimination, governance dynamics and changing livelihoods that have shaped women's land rights. HIV/AIDS increases the vulnerability of an already vulnerable population by increasing the incidence of mortality, marital breakdown, single women hearth-hold relocation, changing the governance of resource access and eroding the labour base. A meaningful understanding of vulnerability has to embrace the other processes that impact on women's ability to gain a livelihood from the land.

Author Curriculum Vitae

Gaynor Gamuchirai Paradza (nee Makura) was born in Marondera Zimbabwe Gaynor attended Dombotombo Primary School, Godfrey Huggins Primary and Marondera High School all in Marondera, Zimbabwe. Gaynor Holds a Bsc (Honors) and Msc (Rural Planning) Degree in Regional and Urban Planning from the University of Zimbabwe. Since leaving University Gaynor worked as a Development Planner for the Zimbabwean Government, Independent consultant on Urban Governance, Women and Local Governance and Gender and Land. Gaynor also lectured at the University of Zimbabwe's Department of Rural and Urban Planning. At the time of publication, Gaynor was employed as a Senior Researcher at the Centre for Policy Studies in Johannesburg where she focuses on local governance policy analysis, service delivery, gender and land tenure, natural resource access and HIV/AIDS in Southern Africa. email address gparadza@hotmail.com

AWLAE

African Women Leaders in Agriculture and the Environment

This thesis is one of a series of AWLAE thesis. It represents the fruits of collaboration between African Women Leaders in Agriculture and the Environment (AWLAE), Winrock International and Wageningen University and Research Centre. AWLAE is a pan-African Program that aims at training women professionals in the fields of agriculture and environment, to redress the existing gap between male and female representation in professions relating to the fields. AWLAE was initiated by Winrock International in 1989. Its headquarters are in Nairobi, Nairobi, Kenya.

Between AWLAE, WI and WUR a project formulated that was submitted for funding to the Minister for Development of Cooperation of the Netherlands Ministry of Foreign Affairs. The goal of the project was to build a cadre of well-trained African women professionals working in agriculture, environment and related sectors to enhance their academic standing and capacity to contribute to gender relevant research and policy-making on the role of women in food systems and the gendered impacts of HIV/AIDS on food security and rural livelihoods in sub-Saharan Africa. In April 2002 the project was granted. The Ministry agreed to fund twenty PhD Scholarships at Wageningen University and the additional leadership-in-change training for twenty women from eleven countries, ranging form East to West and Southern Africa. In June 2002 an agreement was signed between AWLAE, represented by its Regional Director, and the Director of the WUR Social Sciences Group, after which implementation of the project could start. The participating scholars were carefully selected from a large number of applications. The scholarships were widely advertised in relevant media in countries with AWLAE chapters, and the chapters concerned were actively involved in the recruitment and selection of candidates.

The following women participated in the AWLAE scholarship project:

Susana Akrofi	Ghana	Mariame Maiga	Ivory Coast
Namizate Binaté Fofana	Ivory Coast	Lydia Ndirangu	Kenya
Hirut Bekele	Ethiopia	Aifa Fatimata Ndoye Niane	Senegal
Joyce Challe	Tanzania	Faith Nguthi	Kenya
Fatima Dia Sow	Senegal	Carolyne Nombo	Tanzania
Stephanie Duku	Ghana	Regina Ntumngia Nchang	Cameroon
Rose Fagbemissi	Benin	Daisy Onyige	Nigeria
Kidist Gebreselassi	Ethiopia	Gaynor Paradza	Zimbabwe
Monica Karuhanga	Uganda	Corrie du Preez	South
Doris Kakuru	Uganda	Ekaete Udong	Nigeria

www.ingramcontent.com/pod-product-compliance
Lightning Source LLC
Chambersburg PA
CBHW061201240926
51846CB00006B/1084
9789086861460